A Man
After
His Own
Heart

ALSO BY CHARLES SIEBERT

Wickerby:
An Urban Pastoral

Angus: A Novel

A Man
After
His Own
Heart

A TRUE STORY

CROWN

PUBLISHERS

NEW

YORK

Charles
Siebert

Copyright © 2004 by Charles Siebert

Published by Crown Publishers, New York, New York.
Member of the Crown Publishing Group, a division of Random House, Inc.
www.randomhouse.com

CROWN is a trademark and the Crown colophon is a registered trademark of Random House, Inc.

Portions of this book have appeared in slightly different form in the following: Harper's and The New York Times Magazine.

Printed in the United States of America

DESIGN BY BARBARA STURMAN

Library of Congress Cataloging-in-Publication Data
Siebert, Charles.
    A man after his own heart : a true story / Charles Siebert.—1st ed.
Includes index.
    (Hardcover)
    1. Heart—Transplantation—Popular works.   2. Heart—Popular works.
[DNLM: 1. Heart Transplantation—ethics—Personal Narratives.
2. Heart Transplantation—psychology—Personal Narratives.   3. Tissue
Harvesting—ethics—Personal Narratives.   4. Tissue Harvesting—
psychology—Personal Narratives.   WG 169 S266m 2004]   I. Title.
    RD 598.35.T7S54 2004
    617.4'120592—dc21   2003011599

ISBN 0-609-60221-7

First Edition

10 9 8 7 6 5 4 3 2 1

FOR BEX

In memory of
MARION V. AND
CHARLES J. SIEBERT

Acknowledgments

I would like to thank the Columbia-Presbyterian Medical
Center for providing me with the extraordinary access that
made this book possible. Thanks as well go to the Library
of Congress, to the National Institutes of Health, and to
London's Wellcome Institute for the use of their respective
libraries and archives, and to Ronald Carl Marchinkoski
for his invaluable research into long obscure chapters of
my own family history.

A number of authors contributed greatly to my understanding not only of the heart and its history, but of the inherent poetry and mystique of science. There are too many
to list here, but I'd like to thank in particular Dr. Paul Pear-
sall for his work The Heart's Code, F. Gonzalez-Cruzzi, for his
inspired essays about human anatomy; and the paleontolo-
gist Richard Fortey, whose learned and lyrical recreations of
life's earliest origins on earth and the tangible remnants
of those beginnings that we all carry around within us provided me with an endless source of wonder and inspiration.

Thanks to Dr. Robert Michler and associates for the

night of a lifetime, and to Dr. Lameh Fananapazir for the fright of one.

Special thanks to Gerry Marzorati who first suggested I do a book about the heart some six years ago now and has, as both good friend and editor, been by my side on this subject matter from the very beginning; to Susan Prospere and Jeffrey Greene who've been lending ears and insights on this and other subjects for longer than I can remember; and to Francisco Goldman whose longtime friendship and great gifts helped shape this journey before I even knew I'd begun it.

Without the original faith and support of Ann Patty this book would never have gotten started. Without the sustained prodding and oversight of Steve Ross and Doug Pepper it would never have come to fruition.

To Chuck Verrill, many thanks for his loyalty and encouragement over the years, and the quiet wonders of a house called Heath, where the first strains of this work could be heard.

I've been blessed with a big and bighearted family. They know how I feel. To my youngest sister Marion, though, I am particularly grateful in this instance, for her gracious and unstinting love and support during some difficult times.

Finally, I want to thank Dr. Neal Epstein, physician, molecular biologist, philosopher, and good friend, for having the ability to see both the small and the big picture, and the tireless patience to frame them for me in such poignant ways.

A tiny noise is enough to kill you,
the noise of some other heart falling silent . . .

—VICENTE ALEIXANDRE

A Man
After
His Own
Heart

Prologue

Somewhere on this Earth tonight, somewhere, I believe, not very far from me, there is a person whose heart I've touched. A person whose heart I've held in my hand.

One Sunday morning four years ago, in the early winter of 1998, near the conclusion of what was, without question, one of the longest and most peculiar weeks of my life, I reached my right hand into a man's open chest and placed it upon his beating heart.

I have no idea who this person is. I never saw his face. He was lying on an operating-room table at the time, entirely covered but for his split-open chest. I remember, now,

only the look and the feel of his heart, the complete otherness of it against my palm, the beats flush and firm, like the
blows of a fist, resonating the full length of my arm.

It is often at this hour of night that I think of him, the
man whose heart I held: around 4:00 A.M., my "insomnia
hour." When all angles and attitudes of sleep reentry—on
the back, stomach, left side and right; limbs straight, half-
furled or in full-foetal—have been tried and aborted, as
has the book I've been reading, and, finally, the writing
notebook in which I nightly record all those charged
insights that I'll promptly discharge by morning.

This hour when, even as your wife sleeps soundly
beside you, all that's left for you is to prop up the pillow
and wait for your unattended, your unespyed brain—ruth-
less scavenger of sleeplessness—to move in and start making you feel bad about things. Bad enough that you'll soon
be forced to turn over again, and, with one ear or the other
now pressed to the pillow, there have to confront the more
plodding but no less poignant protestations of your heart,
that other essential antagonist in this inherently discomfiting drama of being consciously alive. It isn't much of a
choice we come to at the insomnia hour: whether to listen
to the organ that makes us want to die, or to the one that
can't help reminding us of how soon we will.

Outside my sixth-story windows now, New York City
softly whirs like a huge, idling office machine. Here and
there through the drifting rain, passing cars sound the
two-beat clanks of manhole covers and then splay the
night air, which, like mercury, quickly melds again. Stun-

ning how quiet millions of people can be, how synchronous in the simple quest for sleep.

One, two, three . . . I count the seconds now until the next disturbance comes, and then, in seconds, it does, an ambulance siren threading its way through the neighborhood's streets. I trace the siren to its very last, slow wobbles on the horizon's rim, then turn onto my side, place my ear against the pillow and wait for it, a sound which, prior to that December night four years ago, I would have remained on my back like bound Prometheus and let my brain eat me alive rather than hear.

You often have to shift around a bit to find it, tilt your head slightly off the pillow surface, creating a small hollow between the encased feathers and the ear's cupped inner drum. Sometimes two or three tries are required. You'd think yourself dead if you weren't aware of having that thought. Then, at last, they come: faint, brief snare-beats, like bootsteps through wet snow.

One after the next, I tune into and, because I care to—or because I have cares enough to—I begin to follow them, follow, quietly, in my own heart's tracks, as though trying to skulk away from my brain without its noticing. There is no real escaping, of course. The brain is orchestrating this entire procedure—is propelling, at once, the heart's beats and the pretense of being able to follow after them. I am, we all are, that prototypical prisoner who keeps getting to the apparent edges of his prescribed confines only to be met there by some omnipresent, monitoring orb.

Still, I've found that once the brain is pushed past its

initial resistance to hearing its own thrumming life source, there exists some leeway, a little breathing room wherein the aim simply becomes to keep going, to keep listening to the heart in order to lure the brain away from itself and further down toward the core of the very body it airily oversees and, especially at 4:00 A.M., so deeply disdains.

I've no real sense of how long this all takes. I am not looking at the bedside clock when I am at last about to drift off to sleep. But among the things I do consistently see in this demi-conscious nether realm is an image of myself, rattled and worn, suddenly emerging from a tangled warren of city streets out to the lulling, mist-drifted rim of a great circular fountain.

It is, even for a sleepy, stupefied brain, a fairly predictable simulacrum of the inner biological reality that I am only apparently willing, and yet it does explain why all of my actual, everyday experiences of city fountains and the peaceful trances they induce, are always vaguely marred for me by this nagging sense of redundancy.

And it is precisely at this stage of my sleep reentry—with my brain brought once again to the brink, my consciousness made to see itself again for what it truly is: a mere prismatic arc framed in the mists above its own inner fountain's flow—that he appears beside me: the man whose heart I held.

It's as though I've somehow drawn him there to my side, a feeling similar to the one I'll often have when I'm still lying awake here in bed at night, and the world is silent, and it seems that if I were just to reach my right hand, cupped, into the air and give it a slight turn, I would be able to pull this same man, wherever he is, up out of his sleep.

Or it might be that he has drawn me toward him, has brought me there to his side at the fountain's edge. This too is a scenario I've imagined countless times since that winter night four years ago: I'm walking along a crowded street when my right arm begins to tremble and then, like some built-in divining rod, pull me, irresistibly, in among the passersby, until I'm brought, once again, hand-to-chest with him.

All closed up now, he is wearing a plain gray business suit, and—because it is usually the first one off my mind's dream shelf—has my dead father's face. We stand, silently, side by side, within the fountain's thrum and spray, me all the while trying to recall why we are there together, and how it is that I've come to know this man.

It is, I now understand, a story that has no clear beginning or end, but, like the blood itself, keeps coming back around, full circle.

In each heart, at one time, two motions, the spent blood returning even as the renewed rushes out.

At one time, in each heart—yours, mine, at this very instant—two leanings, two dispositions, two emotions: the urge to go to the very edges of our existence followed by that dire sensation of having gone too far, of being way out on a limb and needing, at all costs, to get back home.

In each heart, at one time, both thrust and thrust's

acceptance, an ongoing, self-contained act of inner coition that at once mimes and moves the outward one to its perfectly mindless redundancy. More and more now I know the outer world to be a recapitulation of our own inner biology.

Outside, the city slumbers along with my brain. I've just doubled back on it, followed my heart's footsteps back around to what I'd taken such pains to escape, found myself standing before some late-night, domino-lit office tower, dead migratory birds strewn at its base.

Where, then, to begin? At what point in the heart's motion to intercede without disrupting that ongoing simultaneity? It has a mind of its own, the heart, for which our minds have yet to find the words.

Chapter

I

I N EACH MIND, AT ONE TIME, PROJECTION AND recollection, thoughts continuously coming, like the blood, full circle, coalescing only in an ever-shifting, evanescent present: December 1998, a Saturday night, a night not unlike this one, me lying here in bed, wide awake, with nowhere to turn. Outside, the entire insomnia cast was waiting in the wings: the whirring city, the light rain, the passing cars, the subsuming silence, the sirens. I know this from experience if not memory.

There were, however, a number of notable anomalies. The time, for one. Just past midnight, only a brief, first round of sleep achieved. This bed, too. It was empty but for me, my wife having gone away to the country because even as I lay here, trying to drift off, I was working. I was "on call," as I had been all that previous week, to accompany a team of surgeons on what is known as a "heart harvest": extracting from the body of the next available brain-dead donor his or her still-living heart and then delivering that heart to a waiting transplant recipient.

Civilians are normally not allowed to attend such a procedure. To be admitted into any operating theater, all nonmedical personnel are required by law to get prior written permission either from the person to be operated upon or from a member of that person's immediate family. In the instance of an organ harvest, however, the former option is obviously out of the question and the latter is rightly deemed unseemly to pursue given the emotionally distressing circumstances that allow such an operation to happen in the first place.

Why I was granted "the privilege," as Dr. Robert Michler, then the head of the Columbia-Presbyterian Medical
Center's Cardiac Surgery Unit, phrased it when I first visited him in his office earlier that fall, is still something of a
mystery to me. I'd originally gone to see Dr. Michler about
the possibility of witnessing a heart transplant, for a book
I'd long been intending to write about the heart. What sort
of a book I wasn't entirely sure. But rather than just lose
myself in the limitless literature, scientific and symbolic,
devoted to the heart over the ages, why not, I asked myself,
go out and meet it firsthand?

The subject has hardly been sidestepped over the ages,
and yet as I tried to explain to Dr. Michler in his office that
day, it somehow seemed to me to be in desperate need of
redress. Not only because of the recent advancements in
biomolecular sciences and cardiosurgical procedures, such
as bypasses and transplants and the latest artificial heart
implants, but also because of what concurrently and somewhat counterintuitively seems to be the heart's marginaliza-
tion in the modern age, its demotion in this, the era of the
brain—the so-called new frontier, with all of its reticular
wonders and as-yet-unraveled mysteries—to second-class-
citizen status: the fully solved and demythologized half of
the human equation.

The heart's current "condition" might best be encapsu-
lated by that phrase "It's just a pump!"—one most often
uttered by physicians who've been driven to utter distraction
by people like me, people who in their not-too-distant, or
not-distant-enough, pasts have rushed one too many times

to hospital emergency rooms, convinced that they're dying in the wake of their "pump's" most recent spot-on imitation of a badly startled, caged canary.

Heart hypochondria—that sudden, all-consuming awareness and fear of the very muscle that moves each of us from one tenuous breath to the next. It has become a fairly common phenomenon by now, especially among my contemporaries, for whom that first late-night dash to the emergency room because of a fast-unraveling skein of skipped beats or a bad bout of indigestion is a kind of rite of passage, a first, close encounter with mortality.

That there is now an ever-growing number of heart hypochondriacs—as I've since heard many a cardiologist complain—is due in part to the greater awareness among doctors and researchers of the heart's functions and malfunctions. This, in turn, has created a kind of half- or shadow knowledge among the lay population, hypochondriacs being notorious for knowing only enough about the source of their fears as they need to perpetuate them.

With a steady barrage of information about the threat and causes of heart attacks (the endless magazine ads and TV spots with freeze-frames of suited businessmen clutching their chests in agony) filtering into the brains of increasingly idle workers, sealed off all day in hushed office towers with little more than the sound of their own heartbeats, it is little wonder that more and more of us are rushing to hospitals with the first phantom chest pains.

I was all of twenty-two when I first got the line about the pump. It was a bitter cold February morning back in the winter of 1978. Having already made three late-night visits to the emergency room in the same week for what the doctor would determine each time to be a mere anxiety attack, he finally ordered me into his office. He began by reminding me that he did have other patients with serious problems, my father among them. He then picked up the plastic desktop model of a human heart before him and, with his pencil, proceeded to poke at and describe the various chambers and valves—atriums, ventricles, aortic and pulmonary, mitral and tricuspid.

As he spoke, I found myself thinking back to those old black-and-white grade-school science films with that same guileless, baritone voiceover: "Your heart beats over 3 billion times in an average lifespan. Like a great city, the body needs a transport system to carry its vital cargoes to and fro. (Cut to overhead shot of some major urban road system.) This network—the circulatory, or cardiovascular, system—has its freeways, underpasses, cloverleafs, subsidiary roads, quiet streets, and back alleys. The total distance they cover, within the confines of the body, is estimated at 60,000 miles. Under the impetus of the heart's pumping action, the blood, with its freight of nutrients, makes continuous round trips . . . the heart pumping so steadily and powerfully that in a single day it pushes the ten pints of blood in the average adult body through more than 1,000 complete circuits, pumping the equivalent of 6,000 quarts of blood a day. . . ."

"You see," my father's doctor said, setting his model heart back down, "a very efficient pump."

That, I remember thinking even at the time, is just another one of those gross oversimplifications that people like him need to make, people whose jobs regularly excerpt them from the flickering surface story we laypeople live by, that thin, light-perfused film of our everyday consciousness. Indeed, it is only when illness causes that film to sputter that we're forced to contemplate its underpinnings, to indulge—as impatiently as we do the tinkerings of a projectionist in a suddenly darkened theater—the equally impatient descriptions and prescriptions of a doctor.

But over the course of the quest that I would begin that long-ago winter day to try to understand the true matter and nature of the heart, I have come to realize that many medical experts actually believe that line about the pump. That their greater knowledge about the workings of the physical organ has led to a diminished appreciation of its abiding metaphysical significance, not, in accordance with age-old conceptions, as the seat of our emotions or purest thoughts, but, as current research is beginning to show, as our brain's subtle antagonist, its emotional and psychological counterpoise.

There is more than mere romance or nostalgia behind the fact that the heart retains its central place in our consciousness as the so-called sun of our body's microcosm, a conception as old as heart symbolism itself, and one that figures prominently in the writings of William Harvey, whose first definitive mapping of the blood's circulation in De Motu Cordis, published in 1628, is regularly cited by present-day, puddle-brained, nostalgic romanticizers of the heart (not to be confused with the ongoingly mystified fanatic such as myself) as the beginning of the rational mind's systematic destruction of the heart's "mythic mystique."

Whereas if people, doctor's especially, would only turn their attention to that same mind's most current scientific revelations about the heart: about chaos theory, for example, and the discovery that the healthiest hearts beat the most erratically; about the fact that people do die of heartbreak; about the heart's "cellular memory," and the secretion of its own brainlike hormones; about the heart muscle's Insect Flight Response: the fact that the same inner-workings of the exquisitely aligned, energy-efficient muscle fibers nature invented way back in evolutionary time in order to power the beating of a fly's wings at 150 times a second, were also borrowed by nature to help power the beating of our hearts; then they would begin to understand that the mystique of this much mused-over muscle is, in fact, so intact, so vital, it now requires even newer and better metaphors in order to be conveyed.

I did my best to rein in such outpourings before Dr. Michler that day, figuring that if there was anyone likely to invoke the "pump," it would be someone who'd already transplanted 250 of them. A tall, broad-shouldered man in his early forties, a crop of straight, neatly combed brown hair framing a round, boyishly handsome face, he'd just performed his most recent transplant an hour before our meeting, a clearly drained yet deeply satisfied chief surgeon, sitting there in his blue OR scrubs, a pair of bright green clogs on his desk-propped feet, a Mozart symphony playing on the stereo—the most essential part, he told me, of his postsurgical wind-down ritual.

It is a curious thing about surgeons. Precisely because of what they've chosen to make a career of—tinkering with and/or taking out and replacing human hearts in this case— you find yourself being deeply mistrustful of the very person in whom you may one day be putting your ultimate trust. And yet there I was, desperately trying to establish Dr. Michler's trust in me in order that I might win his approval to, in essence, play him for a while—to absent myself willfully from that surface story which he was just then trying to climb back into, on the wings of Mozart.

I blathered on for some time about my various heart-related magazine pieces, copies of which I'd brought along as added ammunition. I was, in fact, just getting ready to pull some of them out when Dr. Michler had to field an urgent phone call. It was then that I spotted on his bookshelves one of the very pieces I had in my bag.

It was an essay I'd written for Harper's magazine called "The Rehumanization of the Heart," an essay in which, along with recounting the origins of my own often unhealthy obsession with the heart, I expressed my present convictions about its loss of aspect—at least in Western medical consciousness. At the core of the essay was a look back at what I considered to be the awful apotheosis of the "pump" mentality in the all-too-evident horrors of the first artificial heart implantations that were attempted back in the mid-1980s.

I felt fairly confident that Dr. Michler—he had, after all, agreed to meet with me—was not in league with those physicians who had written letters in response to that piece essentially accusing me of trying to set modern medical science back untold centuries. Theirs, in fact, were the only letters to come for the first two or three excruciating weeks, and then, as if from the distant outskirts of mainstream medicine's hardened affections, one letter after another began to arrive from cardiologists around the country, thanking me, in tones suggesting a defiance of some tacit gag order, for posing some of the very questions about the heart that they themselves had often pondered.

"Oh, yes . . . ," Dr. Michler said upon my describing these wildly divergent responses, clearly combing his brain's over-stacked files for a more exact fix on the essay. Any moment, I thought, he'll be showing me to his office door.

"Yes . . . " he said in his deliberate, sometimes inaudibly airy, perforated voice, "that was you. Very nice. Very nice."

He then stood up, went toward his bookshelves, and pulled from a manila envelope a piece of yellow construction paper. It was a drawing done by one of his recent transplant patients, a six-year-old boy: a simple sticklike figure of a child, his body split down the middle by a thick, dark line. The left side depicted the child with his old heart in his chest, the arm and leg withered, the face sad. The right side portrayed him after surgery, filled out and smiling, bolts of energy radiating from the body.

"I love that picture," Michler said.

For the next half hour he would explain to me, step by

step, the heart-transplant procedure, the vast infrastructure required, and the "incredible drama" as he put it, involved in trying to coordinate the progress of the heart-harvest team with that of the transplant team in order that the healthy harvested heart spends as little time as possible outside of a body, deprived of blood and oxygen—what's known as "ischemic time." Four hours, he said, was the maximum allowable ischemic time before a detached heart began to suffer damage. Dr. Michler then came to the moment when the harvested, ice-packed heart is sewn into the recipient, and the heart-lung bypass machine is switched off in order to allow the recipient's warm blood to flow back to the new heart's still-icy muscle walls.

"The anticipation is there every time," he said softly. "And then the heart starts beating, and all your anxiety dissolves into this incredible sense of satisfaction. The sight of that heart assuming the functions of another person's body. It's the accumulation of so much knowledge, so much science and hard work, and yet each time I witness it, it just boils down purely and simply to a miracle."

I had been taking Dr. Michler's enthusiasm as, among other things, the nod of approval that I was seeking to witness a transplant. But it wasn't until I'd switched off my tape recorder and was about to head out the door that it even occurred to me to pursue the "harvest" aspect of the operation. The very word had struck me at first as being a bit off, and yet the more I reviewed the alternatives afterwards—extraction, excision, removal, reclamation, plucking, salvage—the best I could come up with was "recovery," which still seemed somewhat cold and occlusive compared with "harvest," with its connotations of a fruitful bounty, a life-sustaining yield.

Dr. Michler looked a bit uneasy when I first mentioned the idea of witnessing a harvest. It is the most integral and yet, for reasons I was about to have explained to me for the first time, the most clandestine aspect of the transplant procedure, because of the right-to-privacy laws, the required clearance signatures, and the impropriety of seeking them in such extenuating circumstances.

"Let me think on it," Dr. Michler said.

A couple of weeks passed without a word. I began getting that awful feeling that by asking for a little bit more than I'd originally set out for, I had lost the entire enterprise. Then, early one November afternoon, I received a phone call at my Brooklyn apartment.

"Yes," a woman's voice said. "Dr. Michler calling for Charles Siebert. Please hold."

"Charles? Dr. Michler. I was wondering if you could come see me at my office this afternoon. Three P.M.?"

I got to the hospital—about a forty-five-minute trip, depending upon traffic—ten minutes early. Dr. Michler, his secretary informed me, was running late. Just outside the glass-enclosed oval suite containing Michler's office and those of the other members of the Columbia-Presbyterian cardiosurgical team, there was a waiting area for patients and their families, the usual roundlet of attached seats, magazine-strewn tables, and wan plants. There really is no original way to arrange a waiting room.

A dapperly dressed, late-middle-aged black man sat across from me, his face stuck in a slightly winded pant. "A jazz musician," I'd overheard him say in the course of his conversation with an attendant who'd come by to say hello and update some of the information on his chart. Into his third month of waiting for a new heart, his failing one was currently tethered to the brash, two-beat suck and whoosh of a Left Ventricular Assist Device, or LVAD, sitting at his side.

"Can't wait to be off this thing," he told the attendant, his strained mouth broadening into a smile.

She touched his arm, told him to hold on. In the set of chairs alongside his, a woman sat holding her husband's hand. His ashen complexion and slouched demeanor I recognized from my father's worst days. Directly across from the waiting area was a row of wooden doors, opening onto adjacent conference rooms, those rooms where doctors take family members aside to tell them what they may or may not want to hear.

All the rooms were occupied at that point. Through narrow metal-framed windows on either side of the polished wood doors, those of us in the waiting area could catch only glimpses of the proceedings within, and yet each of us—between flips of a magazine page or shifts in our seats—seemed caught up with trying, trying to discern with furtive stares whether the deep breaths and tearful embraces in one room were of the same cast as those in the room just next to it.

I have, I think, more than the usual aversion to hospitals, my mother having been a ward secretary at one for over thirty years, and all four of my sisters having gone on to become registered nurses. In that fact alone there is, I suppose, a pathology worth studying, the force of one Sicilian mother's nurturing spirit so formidable that somehow all four of her daughters would be swept up by it.

For me it amounted to a nightmarish pastiche of malodorous, pale green foyers and post-shift cafeteria rendezvouses over Salisbury steak and shitty coffee; of overly fawning introductions to vainly dismissive doctors; of endless hospital gossip and TV soap operas and, before long, a few tempestuous trysts of my own with my sisters' various nurse girlfriends and their depressingly adolescent, overcharged sexuality typical of those who daily witness desire's deracination in a brashly lit world of wardroom moans, bedpans, and myriad bodily exudates. The fact that I was now seeking to go to the very core of this world seemed to me indicative either of a deep masochistic strain or of an even deeper need to finally purge myself of all those dreaded associations.

Dr. Michler arrived an hour late, still in his surgical blues and green clogs, and with an even younger doctor in tow, the latter's pens, clamps, and stethoscope jutting from various white jacket pockets. One of the conference room sessions was just breaking up, a procession of dazed faces filing out as Dr. Michler ushered us inside.

"I've got to look in on a patient," he said, turning to leave. "Dr. Rosen will fill you in on the harvest protocol and answer whatever questions you have."

19

I quickly got the distinct impression that Dr. Rosen—
DR. JIM ROSEN was sewn in red script above his breast
pocket—was not so pleased to see me. I was clearly another dirty task dumped by a superior upon one still in the
last year of his residency, and a heart harvest, for all the
"miracles" it entails, is considered to be one more of those
dirty tasks, a physically and emotionally grueling part of a
young surgeon's polishing.

Dr. Rosen would give me the entire rundown that afternoon, step by step, from the initial call to the delivery of the
package: The death vigil. The last-minute, typically middle-
of-the-night dashes to the hospital. And then, depending on
the whereabouts of the recently stricken donor, the journey
either by car or by jet to the side of a body split open upon a
table, the other harvest teams—liver, kidney, lungs, pancreas,
eyes, hands—all watching and waiting anxiously on you,
because in the hierarchy of organ extraction the heart
comes, as it has for so long in our esteem, first.

Rosen's intonation was flat and somber, the tone of a
doctor delivering bad news, the kind they know will knock
you from that surface story, that flickering film we all live
by. There was nothing malicious about this, but rather
something practical, even sensitive. It was his way of eas-
ing me out of the theater and into the heart of the projection room, a place that exists outside of all imagery and
metaphor and artful turns of phrase.

We would need first, he explained, for someone to die.
An organ harvest, however, requires a very specific, subtly
compartmentalized way of dying known as "brain death."

It is as old as any other way, and yet, prior to certain modern technological advancements such as the invention of respirators and the advent of organ transplantation, it was the kind of death that was unsustainable and therefore remained both uncodified and uncultivated.

Sustainable death: a paradoxical nether realm between life and its definitive end. A state in which even though the heart continues to beat on its own, the brain has suffered such extensive and irreversible damage (usually due to some blunt trauma or a brain aneurysm—the sudden rupture of a cranial blood vessel) that it is no longer able to orchestrate the body's vital functions, breathing being paramount. Placed on a respirator, the body continues to function, but like a fully operational factory beneath a darkened, shut-down control room. The flesh still feels warm. The organs remain vital and vigorous.

"We sometimes refer to them as 'heart-beating cadavers,'" Dr. Rosen said.

A neurologist twice conducts a series of tests on such patients, often in front of their family, in order to determine them officially brain-dead: response to pain, sound, light, pupillary or respiratory reflex. And still, Dr. Rosen told me, the body's lifelike appearance can often make the final determination very difficult for a family to embrace.

A recently passed federal law designed to facilitate the process of organ donation obligates a hospital to notify its local organ procurement organization when any patient under seventy-five has been declared brain-dead. When New York Regional Transplant (NYRT), the organ-procurement organization that covers the New York City area, receives a call, it dispatches a representative to talk with the deceased's family about donating their loved one's organs. While a check in the box on the back of a driver's license is an indication of an individual's preference, family consent is still required. Once NYRT receives consent from a family, the organization consults the United Network for Organ Sharing's computerized list to see who in the metropolitan area best matches up with the blood type and size of the donor's organs.

The network uses four criteria in determining who gets the next available heart. The first two determinants are blood type and heart size. A man can get a woman's heart, and vice versa, but a six-foot-four, 300-pound man, for example, could not get by with the heart of a Calvin Klein model. The third determinant is severity of the potential recipient's illness. The fourth is order of listing. "Status 1" means most urgent. If two Status 1 patients are eligible for the same heart, the one who has been longer on the list gets it.

Rosen went on to discuss such matters as our arrival point at Columbia-Presbyterian, the donning of surgical garb in the locker room, the packing of the picnic cooler with buckets of ice, bags of saline solution, and something called cardioplegia fluid, a high-potassium concentrate used to stop the donor's heart just before extraction.

I sat there taking copious notes even as the tape recorder ran on the table between us. I wanted him to know that he had my full attention, still not absolutely certain if or how, for that matter, I would be allowed to come along. Here, I think, was further calculation on Dr. Rosen's part: keeping me in the dark for as long as possible while regaling me with as much detail as he could, all by way of testing my resolve.

"You'll be coming," he finally said, checking his watch and then standing, "dressed as a member of my team. You'll just be like a first-year med student. Our work week starts today. As soon as you leave here, go and get yourself a beeper and then call me with the number. Keep the beeper on you wherever you go. Our call could come at any time. When it does, I'll call you. We'll arrange when and where to meet and then either get in a car or, if the donor is far enough away, hop a Learjet."

He reached into a breast pocket, handed me a card with all his numbers on it, then started toward the door.

"Oh," he said, turning, "if anyone asks when we get to the OR, you're there to observe."

"And if I'm asked to do something?"

"Do it."

Chapter

2

S LEEP IS ESPECIALLY DIFFICULT TO COME BY WHEN you are waiting for someone, anyone, to die. Each night that week I'd lie awake in bed well into the morning hours, staring at that little beeper perched on the desktop across from me, thinking about the millions of people in the rooms stacked all around me, people sleeping or reading or watching TV or making love, and some, at that very moment, dying, dropping off into darkness like a just-doused window light from the city skyline.

Bex, my wife, had already taken off for the country by the time I'd returned home from the hospital with the beeper that first evening. She, too, is a writer, and, given the often discomfiting fact that we both work at home—the constant tiptoeing around the edges of each other's concentration; the stilted lunch-hour conversations conducted under our own tacit gag order on any discussions about each other's work—she felt she'd never be able to focus under the added pall of my ghoulish death vigil.

And neither, of course, was I. I felt as though I had been caught having an affair, and had been left over it, left to ponder my poor behavior and the life that it had now called into serious question. This wasn't an altogether unpleasant experience. At least not at first. It was sort of like traveling in place, the irregularity of my new role putting me at sufficient odds with my normal one to allow me to see it clearly again.

I suddenly saw, for example, that I don't really do anything from day to day, and hadn't been doing anything for quite some time. The fact that I had to borrow the mantle of "heart harvester"—perhaps the utter opposite in the job spectrum to "writer"—in order to recognize this fact, only further underscored for me the level of commitment to, and comfort with, doing nothing that I had achieved over the years, a reality that my ridiculous new beeper, among all the other illuminating anomalies of that week four winters ago, most repeatedly pointed up to me.

It is the only beeper I've ever owned. Indeed, if I have consciously shaped one part of that accrued mass of chance, acquiescence, and arbitrariness that we deign to call our chosen path in life, it is the one that has precluded my needing a beeper. My being beepable. Still, I did my best to conceal this fact at the electronics shop I found on upper Broadway that afternoon just a few blocks from Columbia-Presbyterian.

I stood there nodding my head knowingly at the dizzying array of options and plans that the salesman was offering me, thinking the whole time about how at home my father would have felt here: Charles John Siebert, an electronic parts salesman for National Tel-Tronics for the better part of his adult life, his business suits bearing the same unmistakable smell in the shop that day of Bakelite, the thermosetting plastic used to mold electric parts and circuit boards—a warm, high-pitched, acrid odor, the rear exhaust of your TV's flickering spell.

I eventually decided to limit myself to two essential beeper features: sufficient calling radius and, frivolous as it might seem, color. Following what I'm guessing is an as-yet-unwritten chapter of techno-decorum, the one which declares blue or pink pastel as grossly inappropriate for occasions such as organ harvests, I opted for a more subdued shade. It was a dark maroon unit and yet, in the manner of so many of our mechanisms now, diaphanous, allowing one to see its inner workings. A sweet if somewhat self-conscious phase in our own evolving role as creators: the need to revel in even our crude approximations of biology's inner intricacies.

Once outside the shop door, I got to a phone and called Dr. Rosen with my new number. I then fixed the beeper on my person as inconspicuously as possible: deep in a coat pocket and on the vibrate mode.

It made little difference. I soon discovered that there was no sidestepping my role, perhaps because I felt as though I already had one foot out the door of existence. Wherever I went, no matter how determined I was to go about my usual business, I seemed to inhabit a separate day from the one that everyone else did. At a movie, in a restaurant, or at a bar, I'd peruse the faces around me, wondering who will be next, who—upon the sudden, silent rupture of a blood vessel in their brain, or a distracted misstep at a busy intersection on their way home tonight—will be the next to fall suddenly and forever away from light. Somewhere on the night air, there'd come that warp and whine of an ambulance, and I'd tense for the ensuing vibration in my coat pocket.

I have always been good at waiting. But at an inherently open-ended, unassigned kind of waiting, the kind that readily forgets, if it ever even knew, its objective. It may well be a mental dysfunction. People everywhere these days speak of their "attention deficit disorder," or ADD. Mine would be the lesser known sister affliction: ASD, or "attention surfeit disorder," the most evident symptoms being a disposition to dote on the meaningless. A propensity for prolonged, mouth-agape, apelike awe.

Things like standing on long bank lines, or filling out applications (for anything, for positions, admissions, and accreditation that I've no interest in obtaining), or being told by a mechanic that my car will need another two to three hours on the lift—these have always struck me as peculiarly fruitful occasions, rare opportunities to indulge my affliction while assigning its symptoms the guise of necessity.

All of which would explain, as well, my so-called chosen profession, one that not only best conceals my ASD but somehow seems to be wholly dependent upon it. I'm always amused when people say, eyebrows hoisted, "So, you're a writer!" as though that makes you someone in the vanguard, a person of quick mind and piercing foresight, as opposed to that one figure in life with whom I've always felt the most closely aligned: the lapped runner in a race, the one whose very sluggishness often creates the illusion of being out ahead, while necessarily allowing for a greater absorption of the surrounding details along the way.

But now, with that beeper in tow, all of my daily activities, no matter how pointless or menial, were fraught with urgency. I felt tethered to each passing moment because any one of them could be the one in which my see-through sentinel, my pint-sized prod, would get its first and only jolt in life, sending me, in turn, on the way to a job that, with each passing day, I was beginning to seriously question my desire to perform.

I would never have signed on, of course, if I didn't believe that I had put the worst of my heart paranoia behind me: those thin, evaporative days of constant pulse checks and emergency room visits; the endless, hyperventilative nights, breathing into one of the brown lunch bags I kept in my bedside table drawer. I had more or less willed, or, I should say, embarrassed myself out of this obsessive cycle, having exhausted the limits of physical medicine and no money to even begin testing the psychological.

But perhaps my biggest advance on this front I owe to one pivotal encounter with a whole other kind of professional. A woman named Shirley Solomon, host of *The Shirley Show*, Canada's answer, as far as I could tell from my one near appearance on it, to such American daytime talk-show fixtures as Oprah, Phil Donahue, and Rosie O'Donnell.

I'd received a phone call one morning in the autumn of 1993. It was from one of the show's researchers—Tim, I think, was his name. He was seeking prospective guests for an upcoming broadcast dedicated to ... well, here Tim was very evasive and coy. I would learn from him only later that he had cross-referenced "hypochondriac" on his computer, and it called up the same *Harper's* essay that Dr. Michler had on his office bookshelf that day: the eventual ticket into one of my life's most meaningful encounters serving first as an entrée into one of its most absurd.

I was not entirely clueless in that initial phone conversation with Tim. I made it very clear to him that I would be happy to be part of a show in which I, along with what he said would be a mixed panel of medical professionals, would discuss various aspects of the heart and its prominence in our collective psyches, and so on. I would not, however, be traveling to Toronto for a meager $500 honorarium to appear before a national TV audience as a self-confessed cardiac cripple.

Bex and I took the train from New York to Toronto. Her idea. She thought it would be romantic. We departed on a Friday morning and were to make a three-day weekend of the venture. Her father and stepmother lived in Toronto at the time, and some friends of ours also happened to be in town filming a movie starring—I remember this, perhaps, because it was to be the last film of one of cinema's great "heartthrobs"—Marcello Mastroianni.

The Shirley Show was to tape the following Monday evening, Halloween night, at an old converted movie theater in downtown Toronto. I arrived at the theater in my standard professorial costume: brown tweed sports coat, blue jeans, and button-down shirt. Accompanying me were Bex and a friend, Josh Mostel (Zero's son), whose wife was one of the producers of the Mastroianni film. Bex and Josh let me off at the theater's rear entrance about an hour before showtime, then went to have a quick drink with some members of the film cast and crew before returning for the taping.

"We'll bring back Marcello," Bex said, smiling.

Tim was there to meet me just inside the theater's back door. He looked about eighteen, barely out of high school. He showed me down a narrow hallway to the green room, where, behind the requisite mounded platters of crudites and dip, cheeses and cold cuts, doughnuts and cookies, a young married couple sat on a leather sofa, sniping at each other through tightly clenched teeth.

"Make yourself at home," Tim said. "Shirley will stop by to say hello and give you a brief rundown of the things you'll be discussing."

There was a brief exchange of nods and smiles with what I soon determined from the subject of their bickering to be very recent newlyweds, the husband repeatedly referring to the wife as "a stupid bitch," for ruining their Acapulco honeymoon with her recurring panic attacks, pulse checks, and bag-breathing sessions—one occurring before hundreds of guests in the middle of their hotel lobby.

Soon Shirley appeared. She was big. Not fat. Quite lithe and shapely, in fact. Just big. All of her tight-skirted, made-up features coming at me at once, legs, heels, hips, hands, nails, hair. The hair most prominently: blonde and teased, way up. She greatly resembled the women of my recurring Brooklyn childhood nightmares in which I'm drawn by some inextricable force from my bed along the floor, down two flights of stairs into the basement and the long-nailed, tickling fingers of what I can only guess was my overstoked imagination's re-creation of my mother's daily coffee-klatch friends.

And then Shirley spoke.

"So," she said, sitting down beside me, nestling way too close in her chair. "I'm told you took the train. Too afraid to fly?"

In her right arm, Shirley was holding a clipboard. I could see on the top page an outline of the show's seating plan. It promptly resolved any lingering doubts I might have had about what lay in store. In each of three circles on the right side of the one that signified Shirley were the names of doctors of various disciplines. In the circles on the left of Shirley were only myself and the snarling newlyweds.

"No," I said. "My wife thought it would be romantic."

"Yes," she emoted, breathily, actually touching my knee. "Now I want you to just relax and . . ."

I was promptly given all possible cues to, and various ways of, exposing my tortured soul to the viewing public and a studio audience containing, along with my wife and Josh Mostel, the great Marcello Mastroianni. Shirley then turned to the newlyweds and repeated to them the very things she'd just been telling me. But she did it in such a perfunctory, dismissive way, it suddenly dawned on me that she'd already been through the entire routine with them, and that the bilious display I'd been listening to for the past fifteen minutes had been just that: the two of them eagerly rehearsing for their proverbial moment in the spotlight, which was now, a green-room speaker announced, ten minutes away.

Shirley left us. On the way to and from my makeup session, I had counted at least four bulked-up bouncers guarding various backstage entrances and exits, one milling about in the hallway just outside the green- room door.

"Seven minutes to air."

At the five-minute mark, we were to be ushered stage-ward. I went to the green-room door, and peeked into the hallway. A bouncer was standing there, but with his back to me and, for whatever reason, just then going away from me in the direction of the stage. I turned back to the newlyweds—they were looking at me with genuine fright and disbelief—smiled, and waved goodbye.

My heart was beating torridly. Not out of panic, I think, but pure illicit thrill. I might as well have just robbed a bank. Once out the back door and into the alley-way, it was only about fifty yards or so to the nearest street, but I struggled to get there, feeling that heavy, running-in-a-nightmare tug on my limbs, a bouncer's meaty paw, I was certain, just about to come down on my shoulder. At the alley's edge, I stepped out into the lighted street and found myself directly in the midst of Toronto's annual Hal-loween parade, a benign tide of bobbing, masked faces in which to mask my own.

Tim, the researcher, wound up getting drunk that night with me, Bex, and Josh. Marcello, I'm pleased to say, never made it. The taping, I was told, opened with Shirley looking toward the audience, and asking if a Mrs. Siebert was present. Bex raised her hand.

"Well," Shirley snapped furiously, "your husband bolted."

Bex and Josh had to sit until the first commercial break, staring at my empty chair on stage while I was ducking in and out of Toronto's side streets. I stopped into a bar at one point for a drink, then found a phone booth in an underground parking lot just across from the theater. I tried calling The Shirley Show in hopes of getting someone to page Bex and let her know where to meet me.

I remember standing there, hunched over in the parking lot's sickly yellow light, feeling like a character in a B-movie, Shirley's recorded voice saying something about showtimes and ticket reservations, when a shadow pressed in from behind. I turned. It was Tim, out of breath and sweating heavily. He'd been searching all over downtown Toronto for me. He didn't say a word, just raised his hands and gave me a slight quizzical head tilt. I think he was concerned that I was having a panic attack right there, and that he might have to rush me to a hospital. The calm, measured tone of my greeting, however, seemed to settle him.

"I'm sorry," I said. "The show was just too stupid."

For many years afterwards I would only have to think back to that Halloween night in Toronto: the image of Shirley clutching her clipboard and touching my knee; the battling newlyweds; the specter of my imminent humiliation in front of Marcello Mastroianni—to ward off any encroaching panic attacks.

And yet here I was, sitting up in bed that Saturday night four years ago, staring over at the beeper on my dresser top, Shirley and all the other mental bulwarks that I'd mounted over the years against my paranoia crumbling away, the whole sorry history of my heart and of my father's once again washing over me.

Chapter 3

I T BEGAN WHEN THE TROUBLE DID, WHEN THAT brief grace period, that peacefully pliant union between the brain and the heart called childhood, hit its first real snag. Who, prior to that, recalls much of anything about the heart except, perhaps, for distant, subliminal echoes of its earliest bootsteps beneath the louder, persistent tollings from a mother's ribcage belfry?

But once we're set off on our own, once pulled up into light and air and compelled—after one last, writhing revolt against its coarse emptiness—to take air in, the heart becomes for us much as the sun was in early mythology: phenomenon, a central, redoubtable force, at once life-giving and yet perilous to focus upon.

Many earlier cultures, in fact, conceived of the heart as a throbbing piece of the sun itself, whose power could be reinforced through sacrificial heart offerings—as life-sustaining a harvest in their minds as we were hoping ours to be that winter night four years ago. Forensic anthropologists recently deduced that the Aztecs offered up as many as 20,000 hearts a year, the majority from rival warriors captured in battle. Before cutting out a prisoner's heart, they would tie him to a rope and give him a wooden sword with which to fight once more against his heavily shielded, metal-sword-wielding captors. The longer the bound prisoner survived, the more assured his sacrificers were of the vitality of their imminent harvest.

As for my own childhood recapitulation of mytho-poeic thinking, I was fairly content to keep the heart and the sun throbbing in their separate spheres. And yet there was one heart that I often dreamed of sacrificing. It belonged to the equivalently central and blinding presence of my fourth-grade teacher at St. Augustine's School, Sister Mary Margaret.

Young, tall, slender, lovely, she was named, as she constantly reminded us, after Saint Marguerite-Marie Alacoque, the so-called Sister of the Sacred Heart of Jesus, a seventeenth-century French nun who was beset by a series of intense mystical visions in the form of a transparent, crystalline heart of Jesus, emitting sunlike rays.

According to written accounts of the monk in whom she confided, the heart bore, as well, the wound Jesus incurred on the cross from the spear of a Roman soldier. The heart appeared to Sister Marguerite-Marie bound in its own blood-drenched crown of thorns, Jesus whispering all the while into her ear about the overflowing flames of his love for humanity, so boundless that he was moved to take Marguerite-Marie's own heart out and set it on fire that she might become his love's disciple.

In the years after Sister Marguerite-Marie Alacoque's sudden death, at age forty-three (it is now believed that she suffered a coronary embolism, the prefatory and highly painful bouts of angina being the likely cause of her visions), she would be canonized, assigned her own feast day, the Feast of the Sacred Heart of Jesus, and a First Friday mass in adoration of that heart—a mass that was regularly observed at St. Augustine's church early on the first Friday morning of each school month.

I was so caught up at that time in the aura of this

"Sacred Heart of Jesus"—the image of it emblazoned every-where on prayer cards and schoolroom walls, the heart, in many depictions, radiating from the opened chest of Jesus as he stands staring beatifically with open arms and upturned palms—I believed that the heart itself was kept within the crenellated, satin-lined box at the center of the Catholic altar known as the tabernacle: Jesus' luminous, crystalline heart, pulsating there at the crux of St. Augus-tine's church, and with such enormous power that I feared it would, if I so much as stepped out of my church pew and into the aisle, draw me, like a huge bellows, to the altar.

And then came that Friday morning in the spring of 1963 when fate placed me in the seat at the end of the pew, nearest the aisle. The time for communion arrived. The Catholic Church's rules on receiving communion were quite clear. If you'd eaten anything within three hours of the start of mass, you were not eligible to receive. If you wished to receive, you were to bring your breakfast to school with you to have in class afterward.

Having made sure to eat my breakfast at home that morning, I held fast to my seat. I held to it even as the usher, a huge, menacing eighth-grader, positioned himself by my pew, wanting—I would realize only later—simply for me to let my fellow classmates out.

The usher would eventually have to take hold of the back of my shirt, and pull me into the aisle, whereupon I fell instantly under the spell of the Sacred Heart and began, like some blank-faced automaton, to stride dutifully toward the altar for communion. Of course, the very credulousness that compelled me to imperil my mortal soul that morning would bid me, in turn, to confess my sin within moments of arriving back at our classroom.

I raised my hand. Sister Mary Margaret listened in stunned silence, then called me to her desk. She had me repeat the confession to the class, after which she proceeded to lead my fellow fledgling Catholics in urgent prayer for my soul's salvation.

My own private praying regimen at that age had already achieved a level of nearly pathological intensity and rigor: the nightly sets, three each, of ten Our Father's, Hail Mary's, and Acts of Contrition, with entire sets requiring repetition if I misspoke one word within any one of the individual prayers. But now, with this most mortal sin weighing on me, my every waking moment became a mere backdrop for prayer.

Nothing deterred me. My father took me along one day on a golf outing with a group of prospective buyers for his anonymous little phono jacks and circuit boards. I remember standing in the middle of the green at one point, tending the pin for his putt, the flag stick furtively clamped between my prayer-pressed palms, roars of laughter suddenly rising from the green's fringes, the buyers having spotted me.

I prayed on despite them and my own gathering conviction that I was already, not even nine years into my life, beyond saving, my soul some winged bird lost in a dank bottom corner of Purgatory, cowering under the constant throbbing of that "Sacred Heart of Jesus," the vengeful powers of which I was to witness not a week after my first-Friday transgression against it.

We were on our way, St. Augustine's entire student body, grade by grade, to an afternoon spring recess. I remember filing out the school's side door into a thick whorl of daffodil and sun flash off the nuns' wimples and our own white uniform shirts. Proceeding along to the steeply graded walk that led to the playground, we had gotten halfway up when the boy beside me, I don't recall his name, seemed to just fall off the air, the victim, we'd learn in class the following day, of a fatal heart attack.

I, with extra prodding from Sister Mary Margaret, continued the march with the others up the hill toward that level point where the strictures officially broke down, and everyone spilled out screaming across the blacktop into a fray of swinging jump ropes and punched rubber balls.

As I crested the top of the path, I turned and stole a glance back down through the sun's glare at the rear ambulance doors closing around a small, white, bloodless face that I felt certain the Sacred Heart of Jesus had intended to be mine.

Chapter

4

I DON'T RECALL WHEN EXACTLY I BEGAN TO ABSOLVE both Jesus and myself of any culpability for the subsequent stream of ambulance sirens and flashing red lights that was to course through and, in many ways, demarcate my childhood. You climb down, at some unremarked point or other in your childhood, from that mythic plane of consciousness, and suddenly find yourself just a dazed kid, standing in a suburban side yard of lawn-stranded saplings, your nose stuck in the sun-baked leather of a baseball mitt, watching as your father is being rushed off. Nothing stirs you back then from that airy story of your life, no number of ambulances in sunlight.

He'll be back soon, you're thinking. He'll be out here again, roaming the lawn, as he does every evening upon arriving home from the electronics plant, his jacket reeking of Bakelite, visiting, one by one, with his newly planted saplings: the two maples, red and Japanese; the oak; the white pine; the mountain ash. He'll be here reaching up to touch the branches and buds, as much to pass off his day's tensions to them as to test and receive their vigor. And then later on, before going to bed, he'll be going, with the regularity of the rounds made by his electronics plant's night watchman, from room to room, pressing his lips to each of our foreheads.

The first ambulance to come is the one that I remember best. Perhaps because I wasn't there. Because in each heart, at one time, is the arc of two stories, your own and that of all hearts, mine and my father's beginning on that

particular morning what turned out to be a mutual strug-
gle for more light and air.

I was away in the Adirondack Mountains with a troop
of young Explorer Scouts. We were off on what had been
billed as a regular early-morning hike. Then, some five
miles into the Adirondack forest's cool, piny umbrage, I
suddenly saw the heads far in front of our single-file procession disappearing, dropping down into the earth.

There was a small opening at the base of a great pine,
as though the tree had been slightly uprooted by wind. I
slipped down through it, then made about an eight-foot
drop to a small ledge, sheer rock face on my right, cold
moist earth on the left. From the ledge I could see only
flashlights below.

I edged down toward them through another opening
in the rock, emerging into what we would all soon discover
was the first and largest in a long series of ever-narrowing,
water-filled caverns, the walls slowly closing in around us,
the cold water rising, our hearts pounding, our breath
going pinched and shallow and fast.

And back home in the suburbs. And back home in the
brand-new, sparkly-roofed, Westchester colonial for which
my father had decided to desert Flatlands, Brooklyn, and
mortgage everything, it is a Saturday morning in the summer of 1967, and everywhere there is the plaintive wail of
chain saws and lawnmowers. My father is just coming
downstairs, whistling through wafts of Aqua Velva, eager
to get outside and tend to his corner lot at the juncture of

Orchard Road and Apple Lane, a few old, gnarled refugees from the orchard that our Chilmark development displaced, clawing at the day's too-wide air.

He was, as I remember him at the age of forty-seven, vigorous enough, his life's only major illnesses having been the bout of rheumatic fever he suffered in 1922, at age two, growing up on Manhattan's Lower East Side, and then the near-fatal attack of malaria he would suffer in 1944 while fighting the Japanese in the jungles of New Guinea.

"Doc says I sweated out the bug," he used to say to me in that singularly foursquare Americanese spoken by those of his generation. And while I've since learned there is no such thing as "sweating out the bug," he never did, for whatever reason, suffer the relapses that commonly plague victims of malaria. He got back on his feet in time for the capture of New Guinea—a turning point in the Pacific campaign to which my father, a staff sergeant in the 15th Weather Squadron, contributed by predicting impending storm systems over the region.

Shortly thereafter, in the course of a subsequent two-month leave in Sydney, he was to meet the young, auburn-haired Australian heiress who would nearly eclipse my heart's journey before it ever got started. I found a photograph of her once in an old attic-bound shoe box: the two of them standing shoulder to shoulder in tennis whites on her family estate just outside Sydney.

I have, I suppose, my father's parochialness to thank

for my own and my brothers' and sisters' existence, the engagement to the heiress having been broken off at the last moment because of her refusal (quite understandable, given her accustomed station in life) to indulge my father's homesickness for the pinched, sky-pressed row houses of Flatlands, Brooklyn, to which his father—my grandfather, Charles the First—had moved from Manhattan just a few years before the war.

It was in 1947, in the course of a Saturday beach outing at Riis Park in Queens, that my father was to meet my heart's coauthor, my mother, Marion Valle. Their fate and, by extension, mine and that of my siblings, was, according to family lore, sealed after only three dates by the following telegram sent by my mother from Baltimore, where she'd gone to visit her scores of Sicilian relatives rather than grant my father's request for a fourth date: DEAR CHARLES: HAVING A WONDERFUL TIME (STOP) SHOULD BE HOME IN TWO DAYS (STOP) THINKING OF YOU. CAN'T (STOP). Married six months later, they would go on to raise a perfectly symmetrical—from my middle child's vantage point, anyway—crop of seven, two girls and a boy before me, two girls and a boy after.

They are all there, that summer morning in 1967, seated around the kitchen table. My mother has laid out the regular Saturday breakfast, a typically heedless array of eggs, bacon, sausages, rolls, butter, mounds of doughnuts, and pots of coffee, my father's Chesterfields and beanbag ashtray set there beside his place at the head of the table: a veritable still life of a time before warning labels, when outsized families in Vista Cruisers hurtled headlong and seatbeltless down highways, and no one spoke much of things like nicotine and caffeine and high cholesterol, none of which, we'd later learn, had any causal relationship to my father's heart condition anyway.

He's at the table, savoring the post-meal, last-cup-of-coffee cigarette, his favorite smoke he always said, not feeling so well now, telling himself that the spell of chest-tightness and breathless exhaustion he'd felt on a recent sales trip from merely combing his hair was nothing, and then suddenly he's slumping forward and pale, because his heart, we would learn much later—many years, in fact, after his death in 1980—had faulty fly-wing muscle fibers in it. Was inwardly flying apart.

A congenital disease of the heart muscle itself, no one understood it at that time. No one knew then about the new poetry of the heart, about things like the Insect Flight Response, and the fact that we are, each of us, briefly assembled borrowings from evolution's other, long-ago inspired DNA arrangements. "Idiopathic hypertrophic cardiomyopathy" the doctors called the disease at the time, idiopathic meaning of unknown cause or origin; myopathy referring to any disease of muscle or muscle tissue; hypertrophic indicating a gradual, inexorable thickening of the heart walls. My father's heart, in an ongoing effort to contain its own errant flight, was working too hard, becoming too thick, pumping with ever-decreasing efficiency.

The sirens came, and now the saplings and the apple trees were slipping by sideways above him along with his concerns: seven, and a wife, and the simple pressing question of whether he'd see them again, his chest tightening, his lungs filling, his breathing fast and thin.

And high up in the Adirondacks, and far within its cold, moist earth, silence is giving way now to screams as the walls of the last cave room close overhead and then dive beneath the waterline, only the faintest tendrils of submerged daylight suggesting the way out, the water rising, my chest tightening as I try to force in one deep breath for the long, frenzied, wall-grabbing swim, and someone's waiting hands, pulling me back up into the coarseness of air and the news of him.

I was put on a bus from the town of Brant Lake to Ossining. A neighbor picked me up at the bus station, drove me to the hospital in nearby Sleepy Hollow. They'd gotten him there in time, although among the many doctors clueless then as to the true nature of the trouble, the crudest would advise my mother, "Even if he lives, he'll be a vegetable," a prognosis that a second opinion and an inspired cocktail of the then available heart drugs would allow my father to defy to the tune of thirteen more relatively good years.

I remember him sitting up in his bed when I walked in. He could sense my impatience, seemed deeply ashamed of what he deemed to be my rightful disappointment in him. It can kill a dying heart even faster, disappointment.

That, I've learned, is one of the crueler truths about being us: the only arrangement of life's common clay that continually argues with and wears down that arrangement, and with no one part of it more assiduously than we do our own hearts.

Chapter 5

IN THE ENSUING YEARS BETWEEN MY FATHER'S first near-fatal bout of arrhythmia and his final one—years that happened to coincide with my own adolescence and that invariably tortuous process for a son of trying to sift a "self" from the grist of his father's autobiography—it was the status of his heart by which I marked and now remember the pivotal issues and events of those days.

Would it withstand, I wondered, the lengthened commutes to the electronics plant that his quest for a piece of the suburban idyll had created, to say nothing of the added burden of mounting debt. Will it, I asked myself in the months leading up to the 1968 presidential election, take again the kind of pounding it did back in '64, the year that my father (the only son of second-generation German immigrants and, like his father Charles before him, a deeply patriotic, rugged-individualist Republican who firmly believed Franklin Delano Roosevelt and his New Deal democrats had set the country on the road to ruin) waged his far lonelier crusade in our largely liberal enclave of Westchester County on behalf of Barry Goldwater.

I can still see him standing at the kitchen phone, white shirtsleeves rolled up past his hefty forearms, the Westchester white pages opened before him, another lit Chesterfield wafting smoke up over the finely etched front eave of his already graying crewcut, his face wincing as one caller after the next slammed the phone in it.

With Goldwater's defeat and the slow but steady handover by the U.S. government of the electronics industry to the very nation my father nearly lost his life fighting against—a particularly bitter irony for the malaria-stricken staff sergeant—I watched his disappointment steadily mount.

And then came the evening in late August of 1969, when my older brother Robert, an A-student and star athlete at Archbishop Stepinac High School, came home for a visit before the start of his sophomore year at Holy Cross College in Worcester, Massachusetts. A Naval ROTC officer candidate on a full four-year scholarship, his appearance alone as he came in the front door conveyed the bad tidings.

It was as if the perfectly etched photograph of him in full uniform that my mother had hanging in the den had been blurred beyond recognition by a spilled cup of coffee. His once-shaven head had already grown into a full, curly brown crop. His sideburns had spread down below his earlobes. He had on faded blue jeans and a dark blue sweater with one pale wavy white stripe across the breast. He may as well have appeared on the front lawn with a guitar singing Pete Seeger songs.

Still, it wasn't until the family sat down at the kitchen table that evening for dinner that the obvious was officially declared. Interrupting an already unusually muted round of dinnertime chitchat, my brother announced that he was giving up the remainder of his four-year, full-expenses-paid ROTC fellowship and would be joining a radical campus organization in protest against the Vietnam War.

I recall no shouting or table-slamming. Things, in fact,

went very quiet. My mother kept whispering "Jesus, Mary, and Joseph" over and over. A freshman in high school at the time, I sat there at once in awe of my brother and terrified for him, wondering all the while how he knew he could do such a thing without literally killing my father, delivering to his heart the definitive blow.

I looked over at my father. His complexion was horribly close to the salt and pepper of his crewcut. Then, in a very stern but measured tone of voice, he spoke. He informed my brother that should he go through with his plans, he would be completely cut off financially and fully responsible for his own tuition, terms from which my father would never back down. He then rose from the table and went outside to be with his saplings.

Three years later—years that, even with Nixon's victory in 1968, would witness my father's unlikely transformation from a Goldwater hawk into an ardent antiwar Republican—he would watch with great pride (once he'd gotten past the socialist historian Michael Harrington's "bleeding heart" commencement speech) Robert's graduation from Holy Cross in June of 1971.

But later that same year, with rising Westchester property taxes and the continuing demise of America's electronics industry in the shadow of an ever-burgeoning Japanese market, my father was finally forced to abandon his sapling-strewn plot at the corner of Orchard Road and Apple Lane and accept a lesser post as assistant sales manager at one of National Tel-Tronic's subsidiary plants in the tiny northwest Pennsylvania town of Meadville, mid-way between Pittsburgh and nowhere.

In the middle of my senior year by then, I remember my father coming downstairs one evening to my recently established, surburban-ennui-ridden-teenager's basement bedroom to tell me the news. He didn't lord it over me—didn't, as much as I suspect he may have wanted to at that time, give the hard-facts-of-life speech, the one that would have invariably ended with the "so I suggest you start packing your bags, mister" line. In fact, he more or less asked me if it would be okay if we moved.

Perhaps the most troubling aspect of having a father with such a frail heart is the prohibitive fear it induces in a child of being the one who'll finally do the thing that will break it. There I was at the very juncture when a son is biologically predisposed to beating up a bit on a father's heart, having to tiptoe around his instead, as though it were the very one in Sister Marguerite-Marie's vision, but without the blood and those powerful, pulsating rays: just the fragile, hollowed-out crystal.

I took a breath and told him no, told him that I re-fused to go to Meadville. Hardly the stuff of Greek trag-edy, and yet it did serve to widen what had already become a definitive divide between us. I explained to him that I wanted to finish out my senior year in the same school. He, of course, knew that it had mostly to do with Eliza-beth, the girlfriend he didn't much care for, the eldest daughter of an upper-middle-class, liberal Jewish family in Ossining, her mother a former opera singer, her father a longtime broker at Merrill Lynch, their home a lovely old red brick manse alongside the Maryknoll Center perched high on a hill overlooking the Hudson River.

In short, my sentimental education: book-filled living room shelves and classical music on the stereo; evening cocktails over obscure cheeses and an expansive river view from the house's high, bough-level back balcony; fine bottles of red wine from the basement cave stoking dinner's witty repartee; season tickets to the New York Philharmonic; summers in the Hamptons or Sarasota; the whole of it, right down to my first ever avocado in the post-dinner salad, filling me—in my father's eyes, and I suspect my own if could I see myself now through them—with all manner of pretense and snobbery.

Of course, my regular 3:00 A.M. climbs through our house's basement window, after yet another late-night tryst off in the converted maid's quarters that served as Elizabeth's private bedroom, did little to improve my father's impression either of her or her family. And had he ever learned that Elizabeth's mother was the one who had taken her to the doctor to get her birth control when we first began dating, I'm sure he would have pressed charges.

In the end, it was to Elizabeth, and my growing infatuation with her and all that she seemed to represent, that my father ascribed what he saw at the time as my increasingly aberrant and headstrong behavior. My sudden abandonment, for example, of a potential career in professional baseball, quitting the high school team and passing up the tryout that had been offered me the summer after my junior year by a Pittsburgh Pirates scout who happened to spot me in a batting cage up in Lake George (where my family summered), my father pumping in more and more quarters as he regaled the scout with stories of my additional prowess as a deft defensive catcher.

"Catchers, Chuck . . ." he was forever telling me. "The pros are always looking for good defensive catchers."

It was upon Elizabeth, however indirectly, that he placed the blame for my abject dismissal of all college football scholarships in the wake of countless entreaties and overtures from college football scouts: the fifty-yard-line seats at New York Giants football games; the all-expenses-paid recruitment weekends to various Ivy League campuses; the constant calls from my football coach to my teachers to excuse me from class so that I could come down to the gym to meet yet another recruiter.

My father's heart was to be buoyed briefly by the arrival of acceptance letters from Holy Cross and the University of Notre Dame, places to which I must have applied only to appease him. But then came my own dinner-table bombshell, a resounding dud, I knew, on every front, compared with the one my older brother had dropped three years earlier, but I went through the motions anyway. This, I've learned, is typical of a middle child, the sense of your own exception from history and from the responsibility of even having to make any, because it has all been played out before you once already and when the stakes were far greater and the consequences truly dire.

I stood up before my by then decidedly worn-down, shock-inured father and announced that I had decided to attend a small liberal-arts college called Hobart, at Geneva in upstate New York, where I intended to study literature and creative writing, a pursuit for which there was so little precedent or suggestion in my family that I'd long been sneaking the poems I'd begun writing early on in high school into my older brother's top drawer for him to find on his trips home from Holy Cross.

My parents did, in the course of my sporadic tenure at Hobart, make one trip to Geneva from Meadville for the first "Parents' Weekend," hoping, I think, to relive the joyous experiences they'd had at the same functions at Holy Cross. They arrived early on a Saturday morning in late October, and knocked on my dorm-room door. My roommate answered. I was passed out drunk at the time under a maple tree on the adjacent campus of Hobart's sister school, William Smith. The following night, in a driving rainstorm, my parents and I sat together in their hotel room watching late-breaking news reports about the Watergate scandal: the firing of special prosecutor Archibald Cox and the resignations of Attorney General Elliot Richardson and Assistant Attorney General William Ruckelshaus, what came to be known as the Saturday Night Massacre, the sure beginning of his man Nixon's demise.

I never did go, needless to say, to Meadville with the family in my senior year, and the disastrous house that my father got swindled into purchasing: a split-level ranch on Azalea Lane, with a leaking swimming pool and a perpetually flooding basement. I moved in, instead, with the family of an Ossining friend to finish out high school, then spent the summer with Elizabeth in the Hamptons.

She'd go off to Grenoble, France, for her first year of college. I'd meet her there at the end of mine, yet another venture my father vehemently opposed, even though I bankrolled it all with the rather hefty paychecks I earned from a backbreaking three-month stint as an assembly-line worker at Meadville Malleable, the town's local iron foundry.

And when, the following year, Elizabeth abruptly informed me that my "education" was over, I dropped out of college entirely and decided to go back to Europe, determined by then to get as far away as I could from frail, and breaking, and broken hearts.

This time, however, both Elizabeth and my father wound up pleading with me not to go: my father for fear that I was completely destroying my chances for a sound education and thus beginning a long, slow, sure slide into a life of pointless waiting; Elizabeth because she needed someone by her now, having learned just a few weeks before my departure that her mother was dying of stomach cancer.

I flew to London, traveled from there to Paris and then to Geneva and Leysin. A long, fruitless search for work in one of the nearby mountain ski resorts finally ended one snowy December evening when, while hitchhiking back to my hostel in Leysin, I was picked up by a short, hyperactive Swiss day-laborer named Serge and his plump, soft-spoken British wife, Celia.

"Serge says . . ." Celia began, her head swiveling as I would see it do for the next sixth months, back and forth between her husband's rapid-fire French and my bewildered, expectant expression in the backseat of their beat-up van, ". . . Serge says he needs someone to help him with chores."

"Chores," I soon learned, meant lumberjacking in the Swiss Alps just north of the Italian border. Serge and Celia put me up in their one-room cottage in the mountain village of Mex, a tiny, sky-stranded hamlet halfway between Mont Blanc and the Matterhorn. My working hours were such that we'd depart each morning before sunrise and return to Mex after darkness fell. I didn't see the village in daylight until my first day off, Sunday church bells knocking me from a bone-tired, marrow-deep sleep. I slipped on my clothes and went out for a stroll, a curious "noonday fog" lifting just in time to spare me a 4,000-foot cliffside plummet through the other clouds passing down below.

The bulk of my workdays were spent high on mountaintops, wielding a scythelike claw known as a tourne-bois, rolling Serge's already felled and trimmed trees into a huge gully dug out of the mountain's side. Those logs would, in turn, slide into and dislodge the others already set within Serge's makeshift ice chute, setting off a beautifully discordant cascade of white snow vapor and loud earthen clunks, winding all the way down to the mountain's base.

Six months later, in better shape than I'd been from all my years of football and baseball training, I put down my chain saw in the middle of a workday—the saw with which, in a moment of typical spaciness, I'd just missed lopping off my right leg—walked over to Serge's van, and told him I was quitting.

He and Celia threw me a little going-away party that night at Mex's local tavern. The following morning I was dropped off with a half-year's wages at the nearest train station and from there began a wanton journey, literally and figuratively, south: from Mex to Marseilles to Barcelona and then Madrid, where I came uncomfortably close to getting picked up for shoplifting peanut butter at a highway *tienda* on the city's outskirts just as a cavalcade of seven black limos came flying past at over 100 mph, one of them containing the aging Generalissimo Francisco Franco. From Madrid it was on to Granada, where, tired of roaming, I decided to go a bit farther south and finally took up residence in a borrowed pup tent at a beachside campground in Málaga.

It was, looking back on it now, from Málaga that I began, however unwittingly, my own most pointed assault on my father's already sorely tried cardiac fly-wing muscle fibers. He must have been somewhat appeased (we never spoke about it) by the fact that in lumberjacking I had at least found myself a good job, was earning an "honest wage" (he spoke this way). But there was to be no correspondence in the months after my quitting, a protracted silence that, our differences aside, I assume had him wondering, and that surely had me.

One hesitates to ascribe to any one time frame or set of circumstances such an amorphous and yet pointed decision as the one to become a writer. It is, perhaps, more an act of recognition than of deciding anyway—the recognition, in my case, of a vaguely illicit comfort even with the overt dis-comfort of always feeling excluded or excerpted from things, as though I were already a little bit dead within the larger, apparently thrilling thrum of being alive.

It is that same feeling of exclusion within inclusion that I'm told the middle child of a large family often feels, that sense of being at a slight remove from life and the subsequent recognition to which that feeling often gives rise of a sort of opportunity there, a little recurring loop in that ongoing film of consciousness that allows you to con-tinually insert other, slightly altered versions of events.

It might even be possible, thinking back on that point in my travels now, to retrieve from it some specific por-tent, portent always being retrieved. It might, for example, have been one particular night in which this recognition I'm speaking of fully came to me, a clear, sleepless autumn night when I wandered away from my hostel in a small vil-lage in the hills outside of Granada. I remember sitting up under the stars, reading—like one of those overly devout tourists who travel to the Himalayas with their copy of The Snow Leopard, or to Mount Kilimanjaro with The Snows of . . . , or to Venice with Death in . . .—reading the Anda-lusian poet Federico García Lorca, when I got perhaps one of my earliest and thus still eminently mutable, even liber-ating intimations of that later brick-in-the-face realization that each of us is entirely alone in the world, a little flyaway balloon in the loosening grip of some drunken, dozing carnival salesman.

It would not have been entirely beyond me then to torment my father in a letter with such intimations. There was a day back in high school when I excitedly read aloud to him and my older brother some overwrought essay I'd just written for English class about the tyranny of the time clock and the dead-eyed, gray-suited worker-drones who daily marched in time to it, entirely unaware, as I read on and on, of the offense that might be taken until my brother cleared his throat, and I looked up and registered this look of utter, blood-drained bewilderment on my father's face.

"Geez," was all he said, "I had no idea . . ."

From Málaga, however, all he began receiving from me after months of silence was a barrage of telegrams and postcards. They were filled with updates, idle observations, and talk of possible future plans, but each, in their own way, thinly disguised appeals for money, the very thing that my father sternly warned me, before I left on this second European adventure, would not be forthcoming should I run out.

For well over a month there was no reply. I'd taken to eating vegetables and rice cooked on a little portable gas stove lent to me at the camping grounds by a retired Australian sea captain who was traveling the world with his wife in their trailer. As the Christmas holiday approached, I decided to take up with a group of British travelers who offered me a seat in their van for a week's journey around

Morocco. Five days later we were on the road from Tangiers to Ceuta for our return ferry to Spain. As we passed through a small seaside village, a boy, who'd somehow gotten separated from a group of villagers that had been crossing the road ahead of us, darted directly in front of our van.

The last we'd see of him before being taken away by the Moroccan police for questioning, he was lying limp and unconscious in his father's arms beneath a roadside acacia tree. Nearby villagers had already begun to crowd in around our van, peering in through the windows. Some began rocking the van slightly, tapping at the glass, signaling to us that the boy had been blinded.

Sensing a greater danger in just sitting still, I slid open the van's back door and stepped out. Everyone quietly parted. I could feel on my back the tensile pinch of their stares as I strode off in search of a telephone to call for help. The ambulance happened to arrive a bit before the police car did, creating just enough of a distraction for me to discard my pipe and bag of hash in some nearby shrubs, before piling into the police car for the drive back to Tangiers.

"Here," I remember thinking as I sat panicked and penniless in Tangier's police headquarters, staring at the goat a local villager had tied to a hallway radiator outside the magistrate's office before going inside to argue whatever might have been his case, "here, at last, is my long-sought severance from all the dreary associations of the recent past: no more standing 'on the gangplanks of suburbs' as García Lorca wrote of them, 'letting blood from the stucco of blueprints'; no more leaky swimming pools on Azalea Lane, and

lawn-stranded saplings, and chain-saw Saturdays, and ambulance sirens; no more Bakelite on gray business suits, and
the slow, steady thickening of a miswired heart."

This, I felt certain, is the scenario that will be that heart's
final undoing. The boy we hit will die of his injuries. The
pipe and the hash will be found and traced back to me. I'll
be convicted as an accessory to murder, publicly stoned, and,
if I live, locked away for decades. A letter will arrive at Azalea
Lane, addressed to my father, scrawled in the blood of his
onetime Pittsburgh Pirate catching prospect, explaining
why, given our respective circumstances, we might want
to consider affecting some quick, compromised version of
those rapprochements that fathers and sons with far more
twisted relationships than our own often do, if they live long
enough, achieve.

Late the following morning, after a night sleeping
on the station's hallway bench, an envoy from the British
Consulate came to police headquarters to say that the boy
had begun to regain both his consciousness and his eye-
sight. There still seemed to be some discrepancies, however, between our collective account of the accident and
those gathered from eyewitnesses at the scene. We might,
the envoy explained, be held there through Christmas and
for some time after that.

We were allowed to leave the station that night on our
own recognizance. Driving to a nearby campsite, we parked
alongside a shoulder-high stone parapet that seemed to
stretch for miles. We walked beside it for a time, then
came to a huge wrought-iron gate that opened onto the

Atlantic Ocean, and passed through and fell asleep there on the beach beneath a wide, crystal dome of stars.

The next morning, back at police headquarters, we were chatting with the British envoy when a tall, dark-skinned figure with a finely etched beard appeared. Wearing a pristine white djellaba and a dark maroon turban, he strode directly into the magistrate's office, followed hastily by the envoy. Ten minutes later the man strode out again, without so much as a glance in our direction. The envoy soon followed. He came over and informed us that the man was a witness to the accident. His account supported ours. We were now free to go.

I have no idea what transpired that day. All I remember is we had to go back to the British consulate to fill out some paperwork. When that was completed, we asked the envoy if we might stop over to the hospital before leaving, to visit with the boy and his parents, offer them our apologies and some sort of gift.

"If I were you," the envoy said, "I'd consider myself extremely fortunate and get out of the country as soon as possible."

Another three weeks passed back in Málaga before I finally received a response to the messages I'd been sending from the American Express office, my most recent only a few days after our release from jail. It was one of my father's signature missives, the single, neatly folded page of unlined, premium-bond stationery, crammed with the perfectly aligned if somewhat florid penmanship that flowed from the same slender, silver-plated Cross pen he used all of his working life:

Dear Chuck,

I am glad things worked out for the best in Morocco and to know that you are well. However, regarding your request(s) for money. You'll remember when you first informed me of your plans to set out on this, the second interruption of your schooling, I told you that if you did so, you'd be entirely on your own, and shouldn't bother to come asking me for money. As much as I'd like to help, I can't possibly go back on my word now. I've already discussed the matter with your older brother. He is in full agreement.

Love,
Dad

Chapter
6

W E HAVE, AS YET, NO MACHINES TO MEASURE the specific physical effects of such things: the burden on the heart, for example, which the very life that a heart allows brings. Still, I feel certain that his suffered far more over the substance of that letter than did mine. Just as I know—because he was, although often a stubborn and a shortsighted man, not a petty or a vengeful one—that his stagnating, ever-thickening heart was in no way reinvigorated or emboldened by the bitter turns that were to soon befall my returning, prodigal one.

The fact, for instance, that in the course of my absence, Elizabeth's only refuge from her grief over her dying mother would be found in my older brother's arms. Or that in the winter of 1978—only a few years since I had fired off to my father from that same Málaga American Express office what I declared to be "my last correspondence" with him, cursing his unyielding principles and forswearing all future associations—I found myself living in his house, off of his money, recovering from radical surgery on the knee I'd blown out in a drunken backyard touch football game during my final year of college. From promising prospect to prodigal parasite, no better than the young Siberian scrub jay, notorious for a phenomenon that scientists call "delayed dispersal": the inability to either leave the nest or to do anything to help it.

And then one day, without warning, your own heart awakens you to your self, to who and what you truly are. I remember the precise moment, my own blinding inner eclipse, the onset of my cardiac Dark Ages. It was that very winter of 1978 in which I first heard the line about the pump. Coldest winter for years. Night. I am in the guest bedroom of my parents' most recent suburban home, this one north of Chicago, the only place my father could land a job after his dismissal from National Tel-Tronics and twenty-five years of combing the Northeast, proffering their phono jacks and tiny circuit boards and all the other anonymous little television parts that eventually get assembled into that recondite whole of which we know only the outward, flickering spell.

I'm sitting up in bed, caught, at one time, between two motions: phrenetic exhaustion. My right leg is in a full cast after the knee surgery, the entire Midwest cast in the waist-deep drifts of another major snowstorm. At home now is only the family's younger half, of which I, as the middle child, feel like the pale top of a broken-off wishbone. From the downstairs den, intermittent laugh-coughs and strains of the Tonight Show band waft upward. The inherent foreignness of family. Genetically foisted friends.

It has been a bad winter all around. Three times within the space of two months my father had to be rushed to the hospital to have the failing rhythms of his sick fly-wing muscle fibers righted again. Midway through my senior year in college by then, I had shown up in early January with my ignominiously blown-out knee, my youthful abhorrence of the apparent dead end that my father's life seemed to have arrived at not so deep as to blind me to the benefits of his health-care plan.

I'm lying awake in bed, a pillow propped behind me.

I'm reading Pablo Neruda's Memoirs, the part in which he recalls riding as a child on the narrow-gauge ballast train that his father used to conduct up through the snowy heights of the Chilean Andes. My attentions begin to drift. I can feel all of my weight now pressing off in the direction of sleep but for one spot of levity at the breastbone, a slight raveling of the heart muscle and then—as though the very train I've been reading about has topped its mountain only to find it has no brakes on the far side—a sudden, furious unraveling at well over 200 beats a minute.

It has a mind of its own, the heart, that our minds are sometimes forced to follow. The fastest known heart rate in nature, roughly twenty beats per second, belongs, strangely enough, to both the hummingbird and the shrew, curious cardiac counterparts—one airborne, the other earthbound—and yet equally skittish habitués, it seems, of their respective realms. And whether I had flitted birdlike, or frantically dug my way out of bed and down the front hallway stairs to reach my parents that night, all I remember is arriving at the den door, draped, breathless and pale, over a pair of crutches, appealing directly now to my couched and heart-sick father, as if to say: "You must know this. You must know what to do."

He moved with light, efficient grace for a man of his heft. He would survive just long enough to see the weddings of the oldest two of his four daughters, and I have a very clear image of him at both, slow-dancing with my mother: curious heart counterparts themselves given her much leaner, and nearly equivalent five feet eight inches. And yet, somehow, when framed within his certain step and the tentative, the taking posture of both his waist-wrapped and his upheld, taking hand, they seemed strangely eloquent equals.

While my mother was getting my father's heart doctor on the phone that night, my father was getting me out to the car. The whole time I kept reviewing in my mind what he had told me of the night he first knew something was wrong with his heart, the same tightness in his chest, the dizziness and shortness of breath that he felt just from combing his hair in some New England hotel room on one of his many sales trips.

He helped to lift my cast-encased leg into the front seat and then he hurried to the driver's side. The air was frigid, thin, and I remember worrying about him even then, in my own hyperventilative ether, knowing how hard it was for him to breathe in the cold, and how extra hard, I now realize, to breathe with the added weight of thinking that he may have passed on his own heart disease to me.

He got me to the emergency room. I was rushed inside. And now our roles were fully reversed: he now left to wait in the brashly lit outer room with the fluttering, high-cornered television, while I lay inside, amid all the other emergencies, all the other wide-eyed, blue-curtained outcasts of their own life stories.

I got poked and prodded and monitored with nodes, attendants all rushing madly about along with my father's heart doctor, who, I would later learn, was genuinely concerned for my young life that night. Not, as it turns out, because of any suspected problems with my heart, which, its sudden impulse to acquaint me with the point of view of the shrew and hummingbird notwithstanding, was determined that night—and after three subsequent late-night visits—to be perfectly normal. He was worried, he would later explain, about the possibility, always present for those in a cast after major surgery, of a loosened blood clot from the operation lodging in the area of the heart.

By the end of my third visit, of course, it was my character that was being called into serious question. By the fourth I was deemed a complete nut, an official—to use the phrase favored by physicians—cardiac cripple. There was to follow the inevitable sit-down session in the doctor's office. I remember him as a very prim, compactly built little man of overbearing self-assurance. He spoke in the most clearcut, clinical terms about my episodes of "auricular tachycardia." I would have to look it up afterwards anyway, go up to one of my sister's rooms and take out the somewhat outdated 1949 *Stedman's Medical Dictionary* that was to be passed down in unbroken succession from eldest to youngest sister for their nursing studies.

Auricular tachycardia was cross-listed with a slew of other archaically named cardiac disorders, curious things like Bony, Hairy, Tiger's and Soldier's Heart. What I had experienced is known as Heart Hurry, "a rapid heartbeat brought on by excess anxiety"; "a form of" the definition continues (and it would take me a long time to appreciate the unintended brilliance behind the juxtaposition of the following two words) "cardiac neurosis." As for my dizziness and shortness of breath, these were, in the ingenious replications of psychogenic disorders, the result of nervous hyperventilation.

I listened patiently that day to the thinly veiled scolding of my father's heart doctor. And then he gave the pump speech. He took hold of the garishly painted, rubberized heart model on his desk. (I have one of my own now, saw it a few years ago in the window of a medical shop on the Rue des Ecoles in Paris while researching my heart book and decided to buy it as a kind of memento of my now "former" heart fixation.) My father's doctor opened the model at the hinge and, with a pencil, began the pointing and pontificating about the different chambers, me all the while transposing his voice with the one from the old grade-school science films.

The week after that session, my father would have another episode of arrhythmia, placing him once again in the intensive care unit, and me, for hours, just outside of his room, conducting a kind of double vigil over my pulse and his. But even after his heart was righted again and he got himself freed from all the machinery monitoring it, I would remain tethered by fear to the beating of mine, listening to it, awaiting the next infidelity.

After a time I found myself becoming nostalgic for the way I thought I remembered it being between my heart and me, a kind of continuum of body and mind, an assumption of soundness, that sense you have as a child of being in such a full, weighty recline within yourself that things like sitting on a porch or swinging in a hammock seem unsettling redundancies. Now there was this strange inner hollow, a chasm and a conversation across it between heart and mind where before there had been none.

My father's cardiologist would make one final attempt that winter to convince me about the pump. He had me hooked up to a device called a Holter Monitor, essentially a large tape recorder that, when wired to your chest and hung over your shoulder like a piece of carry-on luggage, records the behavior of your heart over the course of a day. You're also asked to keep a diary listing the times and brief accounts of your activities so that the poor young intern who has to listen to this less-than-riveting recording will have some referents in what would otherwise be a vast, uncharted sea of heartbeats.

I spent a quiet, reflective day, away from others, not wanting to be seen in public wired to a briefcase. I remember feeling sad the entire time, and a little cheap, as though I'd hired a private detective to spy on my own house, waiting for its main occupant, with whom I'd lived without question for so many years, to betray me.

Many years from now, perhaps, some intrepid anthropologist will stumble upon that tape and my diary in the long-buried rubble of a medical records warehouse. He or she will rush it back to some high-tech decoding lab and, to a steady, sonorous backbeat of lub-dubs, be forced to ponder the following sentient musings of a young Midwestern suburban man from the late twentieth century:

9 A.M. Woke up. Went downstairs. Had a bowl of cereal. Two cups of coffee. Nothing unusual.

10 A.M. Sat in the den reading Pablo Neruda's Memoirs. Again, all normal.

12 P.M.–2 P.M. Got on crutches and went to a nearby park. Everything fine.

This last entry was actually the outgrowth of a rather patronizing and, considering my recently declared aspirations, galling bit of advice that had been given to me by my father's cardiologist as I was leaving his office that day.

"Live a long life," he said. "Go outside. Look around you more. Enjoy nature."

It was an oppressively bleak, flat park with a swing set and a few scattered willows, the grounds surrounded on all sides by ranch houses with perfect lawns and far too much sky above them. Birds kept shifting back and forth above the low rooftops. They reminded me too much of my heart.

The day and my diary went along like this: dinner, reading, a little late-night TV with my parents, and I've often thought since of how ridiculous I must have looked to my father just then, his ghostly white son wired to a tape recorder while watching Johnny Carson and making diary entries about his own heart's responses to it.

Once again, there was nothing out of the ordinary. I eventually excused myself. Hobbled upstairs and fell into a sound sleep, the best I would have in weeks. It may, in fact, have been my heart's most exemplary day ever, and by the end of it I knew that I was on my own, that doctors and their machines could have no part in explaining the heart to me because I had neither a serious problem nor a pump.

I had, in one sense, suffered a very common form of heartbreak: a literal rite of passage into adulthood, when youth's continuum of body and mind is forever broken, and we can only fashion from then on analogies for our heart's behavior.

But the whole story of what happened that winter of 1978, not only in my heart, but between it and my father's, I would be a long time understanding. It is a story that, even as I sat up here alone in this bed that December night four winters ago, the beeper peering in at me now from my dresser top, I could sense was about to come full circle.

Chapter 7

THE CALL CAME SHORTLY AFTER 2:00 A.M. I HAD gone to bed early that night, around 11:00 P.M., exhausted from a week of anxious days and restive nights. I remember falling into a deep sleep and then, at about half past one in the morning, far ahead of schedule, the insomnia hour arrived.

The hour when all at once your eyes, like an uptaken doll's, slide open, and you're looking upon a room, upon a life, that you know will be yours again by morning, and gratefully, even optimistically so, but which at that moment and for however many more it will take you to get back to sleep, seem utterly remote and, by every accounting of your now alien and unforgiving brain, a grave error.

It must be, I've long since decided, a natural biological function, a kind of built-in self-wounding mechanism wherein the brain prompts its own manufacture of those chemicals that will induce morning's illusion of hope and renewal by first scraping itself up against the sides of existence. Tossing your entire life up in the air before you for the nightly rebuke. Forcing you to reconsider, without humor or perspective, all the other possible lives that have been precluded by the one you've chosen, and then, in a final, numbing barrage, exposing that so-called "choice" for the cavalcade of chance and arbitrariness that any life largely is: a by now too-cemented, too-heavy accretion, like the proverbial rolling snowball, around what you know remains at its core a frail matrix of misgiving and doubt.

Why her, you'll suddenly find yourself asking, staring over at your wife of many years, your outstretched hand furled, like a dying leaf, just above the eave of her brow. What is that hand, that overvisited view? Is this what you imagined for it?

One by one, now, your brain begins to call up all the other women you've known, and those you think you would like to have known, and the one you encountered just the other morning in your apartment building's elevator, a new tenant, and lovely, and living right on this floor. It parades them all down that ongoing runway of not-to-be, and yet, in the insomnia hour, the definitive path to a life infinitely more desirable than this one you only think you've chosen.

And here, of course, is just the beginning, the kind of second-guessing one does every other waking minute, a mere prelude to the truly medieval, self-flagellatory procession of woes to follow: professional failings, fruitless pinings, personal shortcomings, both mental and physical, that no amount of will or wishing allows you to overcome. These burdens, in turn, are only mitigated by conjuring those completely beyond your control: a panoply of diseases, with pathologies both slow and swift; perfectly imagined car and plane crashes; and then the truly silly, arbitrary ends, the instances of what we commonly call "plain dumb luck": the loosened building eave that lodged in your uncle's brain years ago while he was walking along a Brooklyn sidewalk; or the story of that guy who lives just up the block from you, someone you occasionally bumped into at the newspaper stand, Mark, you think, is his name.

Mark was walking his dog one morning last month,

when she wriggled off her lead and darted into traffic. He rushed out after her, his view occluded by a parked bus, only to get knocked by an oncoming one into a permanent coma. Someone called an ambulance. The dog walked herself home, strode past the building doorman, through the lobby, and into the elevator. The elevator operator brought her up to her floor, where, with a few scratches and barks at the door, she was greeted by Mark's bemused wife.

On and on it goes, all of it spilling forth like any other involuntary organ secretion, your own brain wielding you about like a bitten-off, still-sentient torso in the jaws of a Great White: global epidemics, poverty and hunger, irreparable cultural dissonance, religious fundamentalist furor, terrorism, bioterrorism, pointlessness, imminent abyss— your only options being either to ride it out or, as would happen that December night four years ago, be mercifully interrupted by a matter of actual urgency.

I watched as my beeper danced and flipped like a stiffened fish off the dresser top, then went to the phone and called Dr. Rosen.

"Hey," he answered, his voice eerily calm, nearly playful. "We've got some work to do."

"Where?" I asked.

"Newark. Sorry. No Learjet."

I had remembered Dr. Rosen telling me earlier in the week that the call was likely to come on a night just like that one, a weekend night, and rainy, when the roads were slick and the chances of car accidents rose exponentially.

"The donor?" I asked.

"A young woman. She died earlier this evening of a brain aneurysm."

As I was preparing myself to leave, I kept conjuring images of her, a "young woman" going about her Saturday night, out dancing, perhaps, or sitting at a dinner table, arguing with her parents; or just getting ready to go out for the evening, poised before a bedroom mirror, putting on her makeup, when her story just stops. One vessel ruptures. One tiny conduit in the whole unknowably vast, complex circuitry of cells and tissue and exchanged chemicals that amounts to each cresting second of consciousness, breaks, and with it the film, the entire spell of being alive.

She's lying on an operating table now, her heart still beating, still pumping blood to and around an inert brain, like water over a streambed boulder. It's as old as any other kind of death, but it's one that we now have the ability to isolate, suspend, in order to reap its briefly viable bounty.

They have long been linked, death and the harvest. Sitting in my study late one sleepless vigil night, I looked up "harvest" in my *Dictionary of Symbols*. There were countless biblical references equating the harvest with the Last Judgment: "The harvest is the end of the world," Matthew 13:39; or Matthew's "parable of the wheat and the tares" in which final judgment pivots on "the essential quality of the fruit borne by the individual."

From there I was referred to the instruments of harvest, the scythe, "the pitiless equalizer" wielded in medieval myth by the Grim Reaper, and the more ancient sickle, held by Saturn, "the old, lame god of time." And yet the further I read, I found myself seizing upon the less emphasized sense of hope and renewal engendered in these ancient symbols, the positive side of their blade's figurative double edge: the life-leveling cut that both yields an immediate bounty and starts the harvest cycle all over again, allowing life to begin anew.

I had this absurd image in my mind of all the world's unharvested organs, passing into the void. Of all the past, dearly departed hearts that have been so direly dispatched over the ages—in Canopic jars, in gilded reliquaries and tufted tabernacles; in the upraised, blood-drenched hands of Aztec priests—toward some imagined afterlife, only to shrivel up and die in this one.

And then it occurred to me that modern science has, as it seems to be doing on so many fronts—morphed the substance of past mythology into real matter. Stationed now along the edges of our days, waiting there just above what the poet Philip Larkin called "the solving emptiness that lies just under all we do," are the organ harvesters, poised to reap the actual "fruit of the individual," to catch what would have otherwise slipped forever into the darkness by way of giving someone else more light. From that point forward, I began to think of myself as a kind of modern-day Grim Reaper. A Grim Reaper with a beeper.

I'd been instructed by Dr. Rosen to meet him at Columbia-Presbyterian in one hour in front of Milstein Hospital's main entrance. I began to think the most practical, menial, mundane thoughts: Should I pack some sort of lunch? What clothes to wear to a heart harvest? Why hadn't I asked Dr. Rosen about clothes? Dress shoes or casual? Day-of-the-funeral thoughts, those simple, mindless measures by which we sometimes cling to life's story.

Thoughts and blood, coming full circle: February 3, 1980, my father's heart fully stopped now, my mother moving about the mourner-crowded kitchen of the suburban Chicago condominium my father had recently purchased, all of their sprawling, child-filled years distilled down to a last, shaggy-carpeted duplex hard by a thrumming, two-lane highway, a huge golfer on the lighted sign outside the living room window, directing passing traffic to the eighteen-hole public course behind the development's man-made pond.

I, against my mother's wishes—"Please," she shouted, "no more surprises!"—have decided to go out back to the pond to test the thickness of the ice. I shovel away a clear pane in the surface snow, and then don an old pair of hockey skates from my days as Ossining's ice-skating counselor, shoveling the town's ponds after school against those late-afternoon, nickel-gray, winter skies, the surrounding house lights just coming forward in the twilight, a housewife approaching now through her backyard to give me a cup of hot chocolate as I build a bonfire on the shore and then walk back out to set the orange cones designating where the ice gets too thin.

The golfer is blinking, mid-swing, above the alligator-toothed outline of the condominium roofs. I begin to glide back and forth within the pond's outer fray of bare willow branches, stealing occasional glances at the distant, multi-headed silhouette in my mother's kitchen window.

Two large coffee urns are already brewing on the counter. The platters of cold cuts have been arranged. There's a ziti in the oven, another one in the fridge. Someone has sent over a ham. We are in mourning. The land of cold cuts and ham and baked ziti.

"All this food!" someone is shouting.

"Don't worry, Marion. It won't go to waste."

"Does everyone have directions back here from the funeral home?"

"Yes. They'll find it."

"Chuck, please, no more sur . . ."

They can hear me now, I'm thinking, the fish nosed deep below in their mud-dark rooms, can hear my skates scudding across the pond's hard, bone-lit sky, sensing how we humans hover just a stilled breath above them, even catching, perhaps, fleeting glints of skate blade, like those stars that suddenly dislodge from cold, clear nights and slide under.

It was just minutes after Dr. Rosen's first call that the phone rang again.

"Hey," he said. "Sit tight. Not sure we have a match yet."

"What do you wear to a heart harvest?"

"Wear? Well, anything will do. You're going to have surgical scrubs over it. I'd put on comfortable shoes, though. A lot of standing around."

I quickly dressed: blue jeans, T-shirt, sweater, cross-trainers. I packed a satchel with notebooks, pens, a tape recorder, a change of shirt, and a couple of granola bars. I grabbed the car keys and set them on the kitchen counter alongside the beeper, which I'd retrieved from the bed-

room, thinking I should take it in case Dr. Rosen wanted
to contact me in the course of my drive to hospital. I then
took a seat in the high, barstool-style kitchen chair where I
often sit at the end of a writing day, and I waited. Fifteen
interminable minutes passed before the phone rang again.

"We're on. See you at the hospital."

I figured Flatbush Avenue—Brooklyn's main thorough-
fare and a direct shot to the Manhattan Bridge—would be
fairly clear at that hour, the heaviest traffic coming from the
opposite direction: Brooklynites returning from their late-
night Manhattan revelries. A thin, misty rain was still falling,
all the city's hard lights and building edges broken and
bleeding across the road surface.

"Worst damn driving conditions," I could hear my
father muttering as I turned onto Flatbush, he with all his
years of driving, all the East Coast sales trips with the sam-
ple parts in the trunk.

"Can't see a damn thing," he'd say, hunched over the
steering wheel, squinting through his glasses the way I now
find myself having to do, one more surrender to the ongoing insurgency of his genes: the morphing of my facial
features into his; of my voice and mannerisms; even of my
disposition, the same mood swings, from dour to dimly
cheerful and back again; the same mounting sense of self-
doubt and cranky, sourceless irritability. There's no way
around such genetic devolutions. No way of avoiding or out-
thinking them. How do you construct out of thoughts—the
effluvium of your own biology—a bulwark against the inner
tugs and tides of that very biology?

Approaching the intersection of Flatbush and DeKalb avenues, I could see through the mist a swarm of flashing, multicolored lights. An accident, was my first thought, another possible organ donor for the continuance of someone else's story line. By the time I realized it was a sobriety check, it was too late to retreat.

At least eight cars were stacked ahead of mine and already a good number behind. All I could think of now was a deeply disgruntled Dr. Rosen, standing out in front of the hospital, finally deciding to go ahead without me, figuring that I must have chickened out at the last minute. I hopped out of the car and approached the nearest policeman.

"Excuse me, officer, would it be . . . ?"

"Did I tell you to get outta that car?"

This, I've learned, is pretty much standard protocol for New York City police officers, something they must teach at the academy level. All questions are answered by them with a question. In front of my apartment building one afternoon, a bit confused about the day-to-day shifts in New York City's alternate-side-of-the-street cleaning regulations, I asked a neighborhood patrolman if that day's announced suspension of such regulations applied to the side of the street my car was on.

"You see a broom-symbol on that sign?"

"Uh, no, but isn't it . . . ?"

"Do you or do you not see a broom?"

There is, I've found, no dodging this query reflex. No way of slanting your intonation, not even with the flattest of declaratives.

"Nice day, officer."

"Did I ask you about this day?"

It didn't occur to me until I was standing there before the cop in the rain that night, that I had nothing with me that would attest to or verify the odd mission I was on—no special press ID, no Columbia-Presbyterian hospital pass. A beeper, a tape recorder, some notebooks and granola bars did not make a particularly strong case for anything.

"Officer, I'm on my way to a heart . . ."

"You wanna fine right now?"

I went back to my car and waited, one more sentient bubble in an IV drip through the hardened, constricted arteries of authority. Staring across the way, I could see through the drifting rain the late-night patrons filing out of Junior's, a few of them bearing the restaurant's signature orange-striped leftover bags and string-tied boxes of cheesecake.

My father's was always the last of the front window booths, the one nestled against the restaurant's back wall by the entrance to the stairs that lead up to the second-floor washrooms. Seated in that same booth, before a rare, an extremely rare, an "I'll-make-you-take-it-back-if-it-isn't-cool-inside" cheeseburger, and an ice cream soda, and a copy of that day's Brooklyn Eagle, he'd commemorate in that booth each of my mother's seven deliveries over at Brooklyn Hospital, just across Flatbush on the north side of DeKalb.

It was a ritual he'd initiated back on February 12, 1948, the night when, as a twenty-seven-year-old first-time father, he stood helplessly by, watching a priest perform extreme unction on my mother as she lay near death from the severe hemorrhaging she'd suffered in the course of giving birth to my eldest sister, Madeline.

To this day my mother says her bleeding was induced by a combination of acute shyness and ignorance, both of which she ascribes to her upbringing in a very traditional, first-generation Sicilian family out in Flatlands, Brooklyn. One of six children living in a narrow brick row house on East 35th Street in Flatlands—the land of immaculately swept stoops, arched front doorways, and tiny, hedge-bound lawns—she and her sisters, Theresa and Rose, were kept in the dark about most matters worldly, repeatedly warned by my stern and perpetually philandering grand-father, Joseph, about the wicked wastefulness for women of pursuits like reading and, needless to say, of discussing anything even remotely to do with sex. My mother was once sternly rebuked by him for merely repeating the word "hysterectomy," some wildly mysterious thing that she'd overheard had befallen the eldest daughter of her next-door neighbor, Rosyln Harmitz.

It was, I assume, left mostly to my father to bring my mother up to speed on things, and yet she'd be completely on her own when it came to bearing his lessons' inevitable fruit. She lay there in the maternity ward on the verge of delivery that February night, so mortified by the screaming and carrying on of all the other postwar mothers around her, she steadfastly refused to vent her pain and embarrass herself in that way, a bit of ill-timed decorousness for which she nearly paid with her life.

Upon receiving last rites, she managed to make what she must have believed at the time would be her two final pronouncements. She first promised the hospital priest that my father, a Protestant, would convert to Catholicism, my father having reneged on his original pre-wedding promise to do so when the monsignor at my mother's parish refused to marry my parents in a church because my father's best man, Arty Rifkin, was a Jew.

She then turned to my father and apologized to him for leaving him alone with their child, just as his mother, Madeline, had done to him and my grandfather, Charles the First, dying of tuberculosis when my father was three years old. He was to have no memory of her, despite my grandfather's many determined, last-ditch efforts to instill some: the weekend train trips to an upstate sanatorium, hurrying with my father to the side of his "choiring angel," the woman my grandfather first met during weekly chorus practice in the rear, upstairs chapel of their Lower East Side church.

No one knows what inner marriage of will and want it was—of mind and heart—that brought my mother back around that night. Neither can anyone make much sense of the fact that she and my father would respond to such a narrow escape by tempting fate six more times. But once my mother's vital signs were stabilized and she had fallen off into a deep, restorative sleep, my father made his way downstairs and out the hospital's front door. He walked to the base of the hospital's cobblestone circular entrance drive, bought a copy of the Brooklyn Eagle at the newsstand on the other side of DeKalb Avenue, and then strolled up the block to Junior's for the first of what would eventually add up to eight celebratory repasts, the final one to mark my mother's successful recovery from her hysterectomy, shortly after the arrival of young Joseph, the seventh and final Siebert, in September of 1960.

My debut, just after midnight, 12:47 A.M., Tuesday, December 14, 1954, the fourth, the middle child, must have seemed to my father a relatively routine affair by then, a night, and a time of night—it suddenly occurred to me, as I sat there in my car, staring over at Junior's wide-slatted front window blinds—not unlike this one: my father seated there in his chosen booth, near closing time, thirty-four years now into the life of his heart, looking out on these same rain-swept streets (I know this because I've read that day's edition of the Brooklyn Eagle on library microfilm), having his meal and his thoughts, lighting up his second-favorite, post-supper, cup-of-coffee smoke in order to further stoke whatever admixture of whim and worry, of hopeful forecast and fuming regret it was that alternately buoyed and burdened my heart's author that night.

This is that one aspect of our biology we'll never quite parse, the one for which there are no machines or measures, the moment-by-moment exchanges between the heart and its motionless, skull-bound counterpart, an ongoing internal conversation that is, after all, all that we are: the sum of an ever-shifting to and fro over the course of a life-time between our thoughts and their chest-bound, multi-chambered mime. Days are too wide and our individual tracks through them too variable and specific for anything but our own hearts to be able to keep pace with them, and, by any life's end, offer the truest record of.

As for that Tuesday, December 14, 1954, a son is natu-rally inclined to register the day of his own birth on the positive side of his father's heart's ledger, an assumption that, in my case, I've long based upon the simple evidence of my name. I, reporting in as the second son, allowed my father to finally right the wrong that he'd committed five years earlier against my grandfather by not naming my older brother Robert after him.

"You are not Charles Junior," my father would repeat-edly remind me throughout my childhood. "Your grand-father, Charles the First, had no middle name. Neither do you. That makes you the second. I am Charles John. That makes me Charles Jr. Don't forget it."

I was instructed to sign all official documents, all let-ters, homework papers, book reports, applications, any-thing asking for a name: Charles Siebert II.

"Just do it," my father warned. "Do it out of respect for your grandfather."

Charles the First, also a salesman—his suits bearing whatever confused essence was given off by the medical and optical equipment and the random jewelry that he spent his life selling—tried equally hard, I'm told, to instill some memory of himself in me, walking me "bowlegged," as my mother put it, every day to feed the ducks in the park just up the block from our Flatlands row house on East 37th Street, the very house that he, along with his brother John and his wife, Mae, all moved into shortly before my father went off to war. But Charles the First's heart was to fly apart as well, when I was three years old, just months shy, as it turns out, of what has since proven to be my memory's furthest reach.

Still, I would obey my father's instructions about honoring his father's name, despite the inevitable ridicule and heartache this invited. Thoughts and blood, full circle: September 1963. St. Augustine's School. Near the start of fourth grade. I am, once again, being summoned to the front of the class by Sister Mary Margaret, this time to be publicly cross-examined by her about the "odd little Roman numeral" adorning the heading in the righthand corner of my homework paper.

I remember, in the course of my pathetic, quavering attempt at an explanation, noticing an irrepressible smile on Sister Mary Margaret's face, she, like her namesake, craving few things more than the sharing of her own self-mortification, the final consummation of which my whimpering presence there before the class that morning only further fueled her passion to achieve.

"Well, Mr. Charles the Second," she began, "as we seem to so enjoy showing off, let's have you do a little ballet now for everyone."

The incident was to incite my father's second great

confrontation with the Catholic Church, to which he did, on the wings of my mother's miraculous recovery that February night back in 1948, finally convert. He took the matter directly to the school's principal and sister superior, Sister Mary Christopher, an absolute Babe Ruthian figure, both in body and visage, with a big, bulging belly and a round, flat, pug-nosed face. The force of her rear-end paddle swipes was known to send the heads of many a bent-over student, myself included, right into the blackboard.

Sister Mary Christopher would promptly set straight Sister Mary Margaret, who, in turn, would set her sinister sights on me for the remainder of that year, making fourth grade a protracted misery, one that was appropriately capped off by the bout of acute appendicitis I suffered during my final religion exam. I had to be rushed to the hospital, where a surgeon got the useless vestige out just before it burst and released within me the deadly venom I'd long been accruing both because of and, in my fantasies, for, Sister Mary Margaret.

Long after I'd left her classroom, she would still find ways to harass me. One of her favorites was to sneak up in the course of a slow dance at one of the school's Friday night "mixers." She'd place her hands between myself and my partner, push us gently outward, and, in her high-pitched, singsong voice, say: "Let's leave room for the Holy Ghost, shall we?"

Justice, one comes to understand, is often not attainable in this life, neither through the outward judicial bodies by which we profess to administer it, nor within the one body that we do briefly possess. It, too, only betrays us in the end, rescinding either all at once or in slow, sadly remarked increments, its own temporary stay against matter's prevailing conviction to run amuck. Each of our lives is, ultimately, a kind of show trial, the set verdict of which we naturally try to forestall with as many arguments and, as evidenced here, digressions as possible.

But what, I began to ask myself that winter night four years ago, staring over at Junior's front window, what of the subtle, inadvertent ways in which we hasten our own trial's conclusion, the ways in which we haplessly argue against ourselves and our own best biological interests? What else, I wondered, beyond some molecularly misfiring fly-wing muscle fibers, might have been winnowing away at my father's heart even as he sat there in his favorite booth that December 14 so many years ago?

They are far too close anyway, fathers, too prominent, to really know them within their lifetimes. Somehow they are better pieced together in retrospect, viewed, as are all historical figures by necessity, as hybrids of fact and fiction, a blend that, especially when it comes to fathers, becomes ever more sound and nuanced as time passes, because only then are you able to observe the full extent of their insidious reemergence in yourself.

And somehow, as I sat in the drifting rain that night, waiting for my sobriety check, thinking of him seated over there at his table near closing time, December 14, 1954, a little more than an hour into the life of my own heart, the image of him that began to emerge most clearly was that of a deeply disgruntled, disappointed man.

That this hadn't occurred to me earlier I can ascribe only to my not having lived long enough to know my own true missteps and regrets. Because, in truth, my father had always been strangely forthcoming to me and my siblings about his, readily sharing the details of his mischances and lost opportunities as though trying at once to get them off his chest and to offer himself up to his own children as a cautionary tale about the burden of accruing too many unlived lives.

The ones that most plagued him have long constituted a cartoonishly condensed preface to those of my own nightly insomnia procession: he decides against his dream of medical school because he doesn't think he's smart enough; he flirts with the idea of animal embryonic research, but flunks out of agricultural college, because in butchering class one day he can't bring himself to bring his mallet down upon the head of a baaing baby lamb. A promising new career in dairy chemistry is nipped in the bud by the war, where his scientific bent lands him in the 15th Weather Squadron. At war's end, however, he abandons, for reasons unclear, his new-found passion for weather and becomes instead an insurance fraud investigator, a job he quickly grows to love because it involves so many hours in the courtroom. Still, his plans to parlay this experience into a full-time law career go forever awry when he succumbs to my mother's entreaties on behalf of her older brother, whose new electronics firm, National Tel-Tronics, is in peril of collapsing because he, after a couple of his own panic attacks at the age of twenty-nine, suddenly believes he's dying of some form of heart disease. My father promptly signs on as his brother-in-law's chief electronic parts salesman, thereby passing up his last, best shot at following his own heart for a lifetime of phono plugs and circuit boards and Bakelite.

In a word, faintheartedness. And while that may have been, in part, an outward symptom of a literally faint heart, I now realize that it was also his heart's added and avoidable burden. That within each of our life's trials we—as he himself, I think, had been trying to say all along—can, in fact, argue with and coax our own biology; can even defraud and deeply disappoint it.

And all at once I could sense him, sitting over there: his broken silhouette through Junior's front window blinds, the look on his face, the feeling in his already quietly failing heart: Tuesday, December 14, 1954—a fairly slow news day, judging by the Brooklyn Eagle's headline, NEW HICCUP ATTACK WEAKENS POPE, Pope Pius XII's third such attack that month.

Most of his time that night Charles Jr. devotes to the front-page stories about the violent rainstorm that, even as he sits there, dragging on his post-meal cigarette, continues to lash Brooklyn with forty-mile-an-hour winds, flooding neighborhood streets, downing power lines, causing a borough-wide rash of electrical blackouts and car accidents.

An avid baseball fan, he turns next to the sports pages,

where, despite being in the depths of the off-season, he scans the headlines for any relevant tidbits, still basking, no doubt, in the afterglow of his beloved New York Giants' unlikely victory just two months earlier over the powerful, heavily favored Cleveland Indians in the 1954 World Series.

He was, as he would later remind me countless times, at Game 1 of that series, was able to see firsthand, along with over 70,000 others at the Polo Grounds that day, "the Catch," as it has long since come to be known: Willie Mays's limb-flailing disappearance into the vast canyon that was the Polo Grounds' center field in order to snag, directly over his head (he looks at the precise moment of the catch like a priest holding aloft the host at the Offertory) a towering line drive off the bat of Vic Wertz. A catch that my father would, each time the same scratchy replay of it popped up on the television screen over the subsequent years, instantly turn away from, wave a hand, and snarl disgustedly, "Aggh, you can't appreciate it from that!"

He drags on his cigarette again, feels deep in his chest the faintest twitter along his heart, as though a fly has just alighted there, cleaning now, first front then back, its spindly uplifted legs. He quickly scans the review by the Eagle sports columnist Tommy Holmes, of the previous night's pro wrestling matches at Madison Square Garden: ". . . Antonino Rocca, the barefoot boy from the Argentine, engaged Don Jonathan, 280 pounds of tall Mormon. But Yukon Eric, who lost most of his ear either to a polar bear or another rassler and cannot afford a belt to replace the clothesline that holds up his Levi's, was the hit of this little number. A full description of all that happened would require the pen of Marcel Proust, at least. . . ."

Directly below Holmes's column, my father's eyes alight on a small eight-line item, a report out of Cincinnati that the Giants' arch rival, the Brooklyn Dodgers, or the "Brooks," as the *Eagle* refers to them, are on the verge of signing the University of Cincinnati's eighteen-year-old freshman southpaw, a former baseball and basketball phenom out of Brooklyn's Lafayette High School named Sandy Koufax.

He quietly registers this move, my new heart mate, and its potential effects on the Giants' prospects for the coming season. Then he shifts his focus one column to the right onto another eight-line item: the announcement of a trade by the Giants of their rookie shortstop Billy Klaus for the Boston Red Sox' Del Wilber, a veteran catcher who played in just twenty-one games in 1954 and batted a lowly .131.

"Fatheads," he mutters, shaking his head, turning the page, where the most prominent headline of the *Eagle*'s arts section declares: GATE OF HELL: JAPANESE PICTURE IN COLOR OPENS AT THE GUILD THEATER.

"Jesus criminy . . ."

Another drag on his cigarette. Along his heart, more flies arrive, lift their front legs and then the back. He checks the local theaters: "MGM's Double-Horror Show" starting tomorrow just around the corner at the Fulton Street Loew's Metropolitan. Joan Crawford as a "Scarfaced She-Devil!" in *A Woman's Face*—"Whatever I am, men made me! Shudder and like it!" This followed by Ingrid Bergman, Lana Turner, and Spencer Tracy in Dr. Jekyll and Mr. Hyde—"When darkness came, strange desires were awakened in the soul of this two-faced man!"

He puts down the paper, lights another Chesterfield, flies perched now everywhere along his heart, all of them aflutter, at 150 beats per second, and yet without taking flight. He feels a slight chill within, turns to stare out in my direction, through the window blinds into the driving rain, wondering how it is exactly that he's come to this point in his life's story; how it is that he could have hurled himself this far—four children now and an inwardly warring heart—down a path lit only by Bakelite.

And across the street and up some flights of hospital stairs. And across the street and far above the blaring horns and smeared lights and pelting fury, you lay alone, a fully severed self now, one more fish out of water, one more hard breath count begun.

I remember because I was only, I was deeply there. Because even then I was, like all the others screaming beside me in the brashly lit ward that night, already a distant outcast of my own heart's inherent entanglements. Of that moment before we even have a heart or a brain or a spine or a spleen. When we are but one cell encompassing the inmost suggestions for these and all the other long-ago encoded intricacies of which we each eventually become a singularly dismissed, spellbound sum.

But whether it is the ultimate genius or jest of our biology that it seems to strand us on the very surface of its makeup, render us little skiffs of consciousness atop the deeper dips and swells of our own inward and ever-shifting sea of cells, we can, in the deadest of calms—in those sudden vacuities through which we sometimes catch ourselves falling and then have to speedily invent a thought by which to wrench ourselves back out—we can, if we choose not to arrest that fall, come to feel some intimation of it, of our own heart's beginning.

Chapter
8

THE RAIN HAD JUST BEGUN TO LET UP BY THE time I'd made my way through the sobriety check. It was just past 3:00 A.M. I had less than twenty minutes now to get across the tip of Lower Manhattan and then straight up the West Side Highway to Columbia-Presbyterian, a distance of only about seven miles, but one greatly distended by traffic lights, slick roads, and the inevitable drunken New Jerseyites fumbling behind their steering wheels for various Hudson River tunnel entrances.

At the end of Flatbush Avenue, I turned left on Tillary Street and then right onto the curved entrance ramp of the Brooklyn Bridge, the roadway winding back around toward Manhattan's multitiered tableau of lights, both fixed and moving: mist-mottled office and apartment towers; the snaking red and white of street traffic; the sprocketed flicker of subway cars through East River bridge trellises; the smooth, soundless slides of helicopters and jets: the city like one complex, multicellular organism, thrumming against the darkness.

It was complexity, we now understand, cellular complexity, that first impelled nature to invent the heart. Dr. Neal Epstein, a molecular biologist and longtime friend, had explained this to me one afternoon earlier that summer, the two of us squeezed into Neal's cluttered, closet-sized office at the rear of his laboratory at the National Institutes of Health in Bethesda, Maryland.

I had just returned from a year of doing heart book research in London—London because it is the home of both the renowned Wellcome Institute of Medicine, with its extensive medical library and historical archives, and, back in the seventeenth century, of Dr. William Harvey, the man credited with finally mapping the circulatory route of our body's blood, and the near-simultaneous motions of the heart that at once receives and propels that blood. It was a bit of inward circumnavigation on Harvey's part, the successful completion of which had, in yet another bit of exquisite simultaneity, eluded mankind for as many centuries as had the circumnavigation of our globe.

Neal I like to think of—though he's uncomfortable with such designations—as a present-day New World explorer, one of many out there now on that paradoxically vast frontier of our own biomolecular makeup. A senior investigator at the National Heart, Lung and Blood Institute, he has devoted much of his research over the past ten years to the workings of the Insect Flight Response and the various ways in which its malfunction in our hearts— often the result of genetic mutations—can lead to afflictions like hypertrophic cardiomyopathy (HCM), the disease that killed my father.

It was, in fact, my interest in HCM's genetic roots that had first brought Neal and me together. Shortly after my return from London, I happened to read—in the sports pages, of all places—about a research protocol being conducted at the National Institutes of Health, involving families with a history of HCM. A few of my father's doctors had mentioned over the years the possibility of his disorder being hereditary. Still, my siblings and I—perhaps because none of us has developed any symptoms—have always been able to conveniently delete bad genes from our own private list of suspected causes, the list that includes our father's childhood bout of rheumatic fever, the malaria in New Guinea, and, of course, his disappointments, their cumulative damage only heightened, I figured, by the added stress of us, the seven children that he and my mother, in their classic postwar Catholic frenzy, saw fit to spawn.

But here, now, in a story about a professional basketball player whose doctors were allowing him to play despite his having HCM—provided that a heart defibrillator be kept under the team bench at all times—was the news of this ongoing NIH study involving members of numerous families from across the country who all shared common genetic mutations definitively linked to the disease. This, needless to say, gave me pause.

I felt as though I were being presented with a very direct though potentially perilous shortcut to solving the longstanding mystery of what had happened between me and my father that Chicago winter of 1978, the night, as I've come to think of it, when I first got divorced from my heart. Here, it seemed, might be the ready-made if somewhat abrupt ending to this story: I go to Washington, dole out a bit of my DNA, and from its twisted thread divine my prescripted destiny. It struck me as yet another instance in which modern science eerily evinces the intuitive accuracy of ancient mythology, the Greek goddesses of destiny: Clotho, who spins out each our our life's thread; Lachesis, who measures it a certain length; and Atropos, who makes the decisive cut.

My initial phone call to the NIH was put through to the co-director of the Institutes' "genetic counseling" office, a young woman named Barbara Biesecker. She confirmed that research was under way in which families were being tested for various genetic mutations related to HCM, and said that while there was no quick, doctor's-office-type test yet available, I, along with my whole family, could be eligible to participate in the study.

"The question," she suddenly paused, "is do you really want to know such information?"

Ms. Biesecker then proceeded to give me a crash course in the newly formed categories and myriad contingencies that the science of genetics will soon have us all sorting through. Were I to join in the protocol, for example, I'd be doing so as a member of the "presymptomatic ill"— healthy people who, as a result of ongoing scientific explorations of the full human genome and of the genetic roots of disease, are now getting unprecedented peeks at the probable future course of their health's decline.

Before long, Ms. Biesecker explained, there will be quick diagnostic tests to detect the presence of, or predisposition to, a broad array of disorders. A routine physical exam or even a home-test kit will tell us what kind of cards we've been dealt, hand us our personal genetic report cards.

Of course, where there are cures or preventive measures, such insight is precious. But for the great majority of the diseases for which we are now discovering the genetic origins, there are no cures, even when it comes to so-called monogenic disorders, where there is a direct one-to-one causal relationship between a single gene and the disease, like cystic fibrosis, and sickle-cell anemia, or the devastating neurological disorder known as Huntington's disease. In her own research, Ms. Biesecker went on to say, she had found that when it comes to being tested for a genetic predisposition to diseases that have no real cures and whose date of onset cannot be predicted, there are basically two types of people: the "want-to-knowers" and the "avoiders."

"Some people," she explained, "even in the absence of being able to alter outcomes, find information of this sort beneficial. It seems the more they know, the more their anxiety level goes down. But there are others who cope by avoiding, who would rather stay hopeful and optimistic and not have the unanswered answered."

I knew instantly which category I fit into. It was the existence of the alternative type that puzzled me, the people who would rather not stay hopeful and optimistic, people who are somehow assuaged by obtaining the most detailed account possible of their own life's dwindling options. They, I decided, must be the same sort of people who have lots of appointments and beepers.

Still, having both the fate of my heart's psyche and of my heart book to consider, I hurriedly began trying to forge in my mind a new, third category—what might be labeled as the "near knower," very much akin to the "lapped runner," it occurred to me, insofar as I was intent on observing and learning as much as I could in this instance without having to concern myself either with the time or the precise manner of my own particular finish.

With this in mind, I asked Ms. Biesecker if I could speak with one of the scientists involved in the HCM research protocol, someone who might better inform my decision about whether or not to drag myself and my brothers and sisters down to Washington and into the dark, coiled recesses of our respective genomes. Ms. Biesecker eventually decided to give me the number of one of the study's directors, a Dr. Lameh Fananapazir.

I let the number sit in my phonebook for over a week. During that time I did raise the matter with a few of my siblings and my mother. They all seemed interested, but more, I think, in the idea of it. The practical application they were happy to leave to me, the one with nothing better to do, with all the time on his hands.

"Look," they all seemed to be saying, "why don't you go down there, and get back to us with what you find out."

After considerable reflection and second-guessing, I made the call. Dr. Fananapazir's secretary took my name and number. He phoned me back within an hour.

"Yes," he intoned, and then waited in silence, that familiar conversational gambit often employed by authority figures to draw the other person out, make you blather on to fill in the void. I readily obliged. It was as though I were standing in front of Sister Mary Margaret all over again, launching now into a plaintive, rambling discourse about my father's condition, about the state of my own and of my family's health, and the reasons why I really didn't think that genes were even an issue in this case.

"Bring with you as much material as you can find regarding your father's condition," the doctor finally interrupted, his voice bearing strong traces of his Iranian heritage. "There is" he said before hanging up, "a fifty-fifty chance you have this disease."

Thoughts and blood: the cave walls, once again, closing in and the cold water rising. We had all just been laughing and chatting, I remember thinking. There was awe as we first passed down into the earth: huge rock rooms and the play of our voices and flashlights off the ice-sweat and the slow-swelling stalagmites.

And then, three or four rooms in, the narrowing silence and the cave walls closing in, and all along my inner heart, now, the flies alighting, their wings aflutter, at 150 beats per second, but going nowhere, millions of stuck, spindly legs, pulling at my heart.

"They've got this all covered," I'm thinking, "the adults, the authorities. They know the way out. He doesn't ever really let go, does he, the sleepy carnival salesman with the fistful of flyaway balloons?"

Behind me, the boy behind me, is screaming now. Where are they taking him? Why am I going ahead so quietly, doing it once again, the dazed, automaton walk up the church aisle, toward the radiating Sacred Heart and the imminent peril of my mortal soul.

I put down the receiver. By "material," I, of course, took Dr. Fananapazir to mean my father's medical records, and

over the course of that next week I quickly set about obtaining these, the logs of his various hospital stays while
living in Ossining and in suburban Chicago, as well as the
files kept by his different doctors over the years, even the
one who, back in that winter of 1978, had lectured me so
patronizingly about the pump. Fortunately, I had only to
deal with his secretary.

And then I decided to broaden greatly my interpre-
tation of Dr. Fananapazir's request. I got in the car one
day and made the hour's drive north from Brooklyn to
Ossining, to which my mother had decided to return a
few years after my father's death. She had found herself
a studio apartment there on the southern edge of town,
with a tree-shaded balcony overlooking the Hudson River,
and a job as a ward secretary in the same Sleepy Hollow
hospital where my father was treated for his earliest bouts
of heart failure. She had resettled, in other words, on
the outskirts of her own past, and now I was asking her
to go to the very heart of it in hopes of learning the
future of my own heart and perhaps those of all of her
children.

Having already explained my mission to her over the
phone—my impending date with the DNA diviners in
their high-priced fortune teller's tent—she seemed to consider it all just bizarre and exotic enough to play along.
She was not, however, interested in going out of her way
to help. My mother may have taken up residence amid the
trappings of her former life, but only to let them ease her
passage into the next one.

"Most of Dad's important papers and keepsakes are in there," she said as soon as I walked in the door, pointing to the narrow, foot-and-a-half-long cardboard box that she had already set out on her front hallway table. "You can go through it all on your own time. Our visits are too precious."

I wasn't able to focus fully upon any of the "material" I'd gathered until I was on the train a few days later from New York to Washington—on the train and sufficiently lost in that dreamy suspension of rhythmic conveyance that trains induce. I was drawn, naturally, to the contents of the box first, even though I had no idea what bearing any of it might have on the impending question of my heart's genetic fate. Shy of finding documentation of a previously unknown childhood ailment, or of a particularly vivid "upheaval of thought," to borrow Proust's phrase, expressed in a letter or diary entry—the literary equivalent of an electrocardiograph needle's sudden spike, indicating some emotionally induced internal cardiac revolt—I was fairly certain I'd find nothing of interest to an NIH researcher, nothing that would either definitively prove or discount the possibility of misaligned fly-wing muscle fibers.

Still, I felt that the records and mementos of my father's daily life could be, in their own way, as reflective of the state of his heart as were the hundreds of pages of charts and EKG blips documenting that heart's slow, steady decline. If nothing else, I was taking that box only for myself and my own peace of mind, as a kind of complement, and possibly even an antidote, to the medical files. I was taking the box, in fact, for the very absence of illness to which its contents seemed to attest, the simple, haphazard details of a life offered up as a defense against the strict determinism of genes.

Among the many curious items that my mother had chosen to keep were my father's canceled checks from the late 1940s, during the first years of their life together on East 37th Street in Flatlands; and a leather pocket case containing business cards from the various Chicago-based companies to which he was, until the very end, trying to sell his electronic parts: Motorola; Fidelitone, Inc.; Halcyon Injection Molding; Die & Mold Corp. I also found a stack of Exec-Mate daily planner books for each month of 1973, every other page filled with my father's handwritten directions for an upcoming sales trip, their very rendering in his always perfect penmanship—the boldly capitalized route numbers standing out like bridge stanchions above the elegant sweep of the lowercase connecting prose—all exuding a sense of true confidence and self-satisfaction, the tacit conviction that these words at least, in a world of vagaries, false leads, and dead ends, actually amounted to something, could really get you somewhere.

The author himself, of course, made a markedly different impression: moody, retiring, deeply modest, not qualities companies typically seek in a salesman. He simply had none of that self-important swagger often adopted by those who sense their relative unimportance. There were any number of them seated around me on the train that day: salesmen, pitchmen, ad men, foremen, the various clerking cogs of free-market capitalism, purposely shouting into cell phones every last detail of their dealings: "Mary? Hi. Yeah, Bill here. Hey, I need you to get hold of . . . Oh, and about that . . ." and on and on, as though to say, "Look at me, I'm delegating."

No, it seems my father's fascination and facility with mundane things like directions long predated the occupation that would most require and ultimately exhaust those traits. I came at one point upon my parents' 1947 "Honeymoon Diary," the daily record of their drive from, as it happens, New York to Washington, D.C., and back home again. They took turns writing entries, my mother one day, my father the next. Page after page, the same pattern obtains: my mother's entries chatty, emotive, aimless, remarking on the weather and the meals and the landscape. My father's: "Took Rt. 40 south for about two hours and then hooked up with Rt. 15 west. Got into Philadelphia near 4:30 P.M. . . ." and so on, all bridge stanchions and lowercase connecting prose.

Tucked deep within the box's dizzying mix of documents was a separate, rubber-banded stack of papers that, upon closer inspection, I realized was my mother's own mini-narrative of her husband's life achievements and various career pursuits. I found his report cards from Brooklyn's P.S. 222, the same school my mother attended. He seems to have been a complete dunce in English but a whiz at earth science. There was a term report from agriculture school in 1939, where he excelled at "Farm Practice" and

"Plant Protection." I found a graduation certificate from the Air Corps's Weather Observers Course in 1942, in which he earned his best grade, 98 percent, both in "Upper Air Observations" and "Plotting Atmospheric Vertical Cross Sections"; and, finally, a diploma denoting his completion of a course in legal jurisprudence.

One of the last items in my mother's private stack was a five-by-seven color photo of my father at an electronics sales convention, manning the National Tel-Tronics exhibition booth. It appears to have been taken sometime in the early 1970s, and I could smell the Bakelite just staring at it. Dressed in his signature gray suit with a white shirt and broad, striped tie, he is seated at a table draped in bright blue pleated cloth, stacks of promotional literature set upon it, a towering lighted triptych-case of electronic parts swung open behind him. On the exhibition booth's far side, a floor-stand ashtray and a pair of metal chairs with white vinyl cushions waited above the showroom's deep green, floral-printed carpet for the approach of a prospective buyer.

"They are suppliers," reads the photo's caption, "who have goods and services to sell in return for money. Charles Siebert, salesman for National Tel-Tronics, which has been furnishing connectors, clips, anode leads, lead assemblies, and terminal strips for the past twenty years."

His left hand propped on his left thigh, his right arm resting across the table, he is both leaning and staring defiantly into the camera (it would be impossible to overstate the depth of the misery in his expression) even as the neatly receding rows of rectangular ceiling lights above him give the impression that he and his entire display are being sucked backward at warp speed into a great void.

And yet, as I sat there on the train that day, digging ever deeper into the past while speeding toward a date with my own destiny, I began to feel a renewed sense of optimism and joy. The by now well-defined distinction between the medical and the metaphoric heart notwithstanding, it suddenly seemed impossible to me that two such decidedly different hearts as my father's and my own could have the same genetic alignment or misalignment, in this case.

Here was a man of earth science and farm practice and unswervingly precise directions: a man who, even when he looked to the heavens, mapped and plotted them. He read, when he did read, newspapers, mostly, and the occasional popular history text, or dime-store detective novel. There was even one in the box: *The Sixth Commandment*, by Lawrence Sanders, with a paper clip at page 313 holding a torn piece of paper upon which my mother had scrawled, "Dad was up to this page the night he died."

I remember, in what turned out to be the last months of his life, when I tried to explain to him—in doubtless sickeningly high-flown and pretentious terms—why I was giving up my well-paying job as a butcher at a New York steakhouse to go study at a creative writing program in Houston, Texas, he listened to me patiently and then very politely, cautiously, posited, "Yes, writing, okay, but why not journalism, sports reporting, something a bit more practical."

We humans are not, I thought to myself, just these will-less passengers of our own biology, racing along pre-set tracks, staring helplessly out the eyelet windows of the self-driven train of our genes. There is an 'I' sitting right here in this seat, an ego, a consciousness, a me: that distinctly aligned singularity which—even though that in itself may well be little more than an evanescent vapor exuded by my body's billions of years of accrued cellular memory—still allows me to look down at that same biological accretion and regard it, assess it, call it a name: Charles Siebert, Charles Siebert "the second"! Not Junior. Not, as my father himself always insisted, after him.

And then, somewhere near the outskirts of Baltimore, I found, pressed up against one side of the box, a brown leather eyeglass case, the glasses still inside. I slid them out. His, of course. I recognized them instantly: horn-rimmed, greenish brown, square-shaped lenses. Clark Kent glasses. The very pair I had just seen clipped inside the front pocket of the white shirt he was wearing in that National Tel-Tronics convention photo.

He only used them for distances, as I recalled. He had to take them off to read. And now my wife Bex's voice was echoing in my head, all the times she'd caught me squinting in the past months to see road signs or the movie screen, pleading with me to get my eyes examined. I set the box down on the seat beside mine, turned the glasses over and over. Then, if only to stop my hands from shaking, I put the glasses on.

It was like the nether side of Cinderella's deliverance via

her glass slipper. Here now was the world in noxiously perfect focus, everything from the premature bald spot atop the cell-phone blithering jerk at the farthest end of my train car, to the eerie white glyph of downtown D.C., jutting above the treetops far outside my train window. Years of defining my "self," my own cellular muck and moil, in direct opposition to his—me as his emotional, intellectual, and spiritual antipode, his separate, far-seeing son—when all the while and literally beneath my eyes, I had been molecularly morphing in the direction of his myopia, of his very own long-ago prescripted shortsightedness.

I felt certain that had I, in some sudden backward transmigration, been made to lie down at that very moment, forty years earlier, in his bed at night and stare out from his middle-aged eyes, I wouldn't have blinked, would have fully recognized and embraced his own insomnia-hour procession of worries and woes.

I had, in essence, become him, had cellularly amounted to him, just in a slightly different guise: a profferer of poems rather than phono plugs, and yet both of them, in the end, just parts, carefully molded and hard-wired for anonymous service in the casting of spells.

Chapter
9

EXITING WASHINGTON'S UNION STATION, MY father's eyeglasses still on, I passed like a ghost along that city's ghostly downtown streets, genes now looming in my mind as large and unyielding as the cold white monuments and office buildings all around me, edifices that somehow exclude the very humanity they mean to both celebrate and serve.

I walked. I kept walking, as though passing physically out of and away from my former self, half hoping all the while that some old business associate of my father's would cross my path—one of those glad-handing, sales-convention colleagues, for example, who must have slipped him the many "joke" business cards he'd accrued over the years, cards that my mother had also, for reasons unfathomable, chosen to save in that box, binding them separately with a rubber band: "Due to the fact that I will be knocking off around noon to play golf," reads one, "I won't have time to call on you again this summer. Please mail your orders to my attention. I will be around again in the fall kissing your ass as usual."

This fellow, whoever he was—someone my father no doubt politely suffered, offering him a big fake belly-laugh by way of preserving his feelings—would bump into me on the street and, mistaking me for my father, be so shocked at how little I/he had aged over the years, drop dead right there on the sidewalk of a heart attack.

I ended up walking all the way to my hotel in George-town that afternoon, a short walk from the Red Line subway service, which, I'd been told by Dr. Fananapazir's secretary, would deliver me directly to NIH the following day. It was early evening by the time I checked in. I unpacked my bag, put my father's glasses back in the box—it, and all of the potentially hopeful connotations of its contents, obviated now by the power of the almighty gene—and then went out and got blind drunk.

I woke some time after noon the following day, packed my father's records—the medical records, at least—in a shoulder satchel, ate lunch at the hotel, and then caught the Red Line to Bethesda, arriving nearly two hours early for my scheduled appointment with Dr. Fananapazir. An "avoider" no longer, I was, in fact, incredibly anxious now to receive final verification of the fate I felt sure I had already glimpsed.

The escalator at the NIH underground stop was the longest and steepest I'd ever seen, dwarfing even London's infamously deep burrows: a good two-minute ride up from the earth's core toward a tiny host of light. It was as though I were in a huge hypodermic needle, observing my own slow injection into what, upon emerging into it, seemed to me a familiar and yet whole other vein of existence.

Somehow my too-honed awareness of the invisible inner minutiae that determine our outward makeup had now put me at complete odds with that makeup and all the other visible arrangements of it around me in the so-called object world. Here was a day like any other, and yet one seemingly on standby: the trees, the buildings, the people, the sky, all remotely tired of themselves, pining now for reunions with their own disassembled particulars.

What we've come to know of those particulars, I kept thinking as I made my way up a slight hill toward the NIH campus, what we've already discovered of existence's unseen reticulum: the ways of cells, molecules, genes, atoms, quarks, neutrinos, photons, gravitons, phonons, and on and on, it all seems to leave us and the other fully formed fixtures of the observable, the already mapped world, out in the lurch: paradoxically plodding apparitions, just as Plato long ago intuited, of our very own unseen, scintillic symmetries.

It took me a good while to find NIH's Clinical Center, but then I had abundant time to kill. Wandering dazed through a vast, lawn-limned complex of tall, austerely etched research buildings, I found myself drawn at one point to a more humble, single-story, flat-roofed brick structure resembling an old army barracks.

It wasn't until I was standing in the front foyer, its walls covered with computer readouts and a long, lighted glass case containing what looked to me like one of those multicolored timelines of the history of the world, that I learned I had stumbled into Lab Building No. 9, a focal point of none other than the Human Genome Project, that much-ballyhooed endeavor—often referred to in the newspapers as the biological equivalent of everything from the Apollo moon missions to the discovery of the Holy Grail—to produce a complete readout of the human biological blueprint:

a detailed map of the precise sequence of the 3 billion or more pairs of chemical bases that constitute the DNA coiled within the chromosomes inside each one of our cells, including, of course, somewhere down along that vast microscopic inscape, those very genes that power both the beating of a fly's wings and of the human heart.

"This is a sequencing laboratory," a high-strung young man whose lab-coat nameplate read DR. ERIC GREEN gruffly explained upon intercepting me in the lobby. "And you are . . . ?"

It did seem alarming, at first, both to Dr. Green and myself, that I had been able to stray directly into the midst of such a vaunted high-tech enterprise. And then, upon second thought, it didn't. I began explaining hurriedly to Dr. Green who I was and why I'd come to NIH, but somehow the air of high-alert security breach had already dissolved around the shared realization that, for all the mystique and grandiose language surrounding this big science expedition—the "mapping" of our "genetic architecture," the compilation of the "Book of Life," etc.—there was precious little about it even to see, no less steal: brashly lit rooms of gently humming Advanced Biosystem sequencing computers, laser scanning clear, pressed gel plates of DNA.

It was as though I'd made my way into some cutting-edge research facility in the heart of Silicon Valley, only to emerge with pockets full of sand, the pages on the walls around me containing endlessly tiresome rows of A's, T's, G's and C's—adenine, thymine, guanine, cytosine—the four bases of DNA, the common clay of all living matter. Those Apollo moon missions and their once-otherworldly images of bubble-suited astronauts stiffly adrift above moondust now struck me now as quaint relics, the last remnants of the old-world mode of exploration, when the explorers could still bodily visit and inhabit their undiscovered terrain.

"I prefer to call it an encyclopedia," a more relaxed Dr. Green said to me as I stared at the multicolored "map" in the lighted case. "We've been mapping chromosome number 7 here for the past five years now. We're only about half done."

Dr. Green, I soon discovered, had that signature hair-trigger garrulousness of the present-day new world explorer, a trait similar to that of a rural shut-in who suddenly finds himself in another person's company. And unless you have the good fortune, as I had later that day, of meeting up with the likes of a Neal Epstein, you invariably end up kicking yourself for offering such an individual the slightest encouragement.

"You know," I remember interrupting him at one point, "for all you read about them, it is never made quite clear what exactly a gene is."

Dr. Green promptly launched into a long, jargon-laced discourse on everything from the origin of the word (it derives from Charles Darwin's now-discredited pangenesis theory about how acquired traits are passed from one generation to the next) to the fact that genes, random isolated stretches of DNA totaling no more than 5 percent of our entire genome, seem to be the only parts of our DNA that actually do anything, that synthesize or "code for" as Dr. Green put it, the variously arranged proteins that make up each and every part of us, be it a knuckle or a heart.

I was then given a dizzying, Chinese-box-like exposition of genes themselves. Genes have, it seems, like the larger strands along which they occur, their own "noncoding" or "junk" regions called "introns," and their protein-coding regions called "exons," including special DNA sequences known as "control elements" that modulate the duration, amplitude, and area—heart, liver, brain, etc.—of protein expression. Tucked within those exons, meanwhile, are the "codons," the tiny wielders of the amino acids that make up the different proteins that make us.

Somewhere in the course of this mind-numbing ramble, I was reminded again of a beguiling truth about modern science: The closer it seems to bring us to the very essence of our and of all being, the less adequate words are for capturing that experience. It is as if, after all the mythic monsters and curses we have conjured over the centuries as the God-appointed guardians of ultimate truths—the fire-breathing dragons, the sudden earthquake fissures, the snake-filled pits, the skin-flaying potions, and so on, all those obstacles we've imagined coming between ourselves and the attainment of our myriad Holy Grails—the most formidable one, the one that the gods themselves, whoever they might be, have been putting the most stock in, is our own inarticulateness.

I remember asking Dr. Green at one point who it was specifically that was being mapped. Whose genome was chosen to be the paradigm of all humanity? Looking a bit impatient with me now, he replied that DNA from many different people was being used, and that it really didn't matter who it came from, because everyone on the planet is 99.9 percent the same, DNA-wise—a revelation I wasn't quite sure whether to be hurt or heartened by.

"We are getting a crude readout," he said, practically showing me to the door now, "of a consensus human. It's like we're just getting the page numbers of the book down. Then, as we begin to refine our map and fill in the details of each page, we'll be able to develop things like quick, precise tests for various diseases. A diagnostician will be able to turn, let's say, to page 2,500 of chapter 10 and, comparing it to your own genomic readout, see how a tiny misspelling, a typographical error in your DNA, might give you and the rest of your family hypertrophic cardiomyopathy, for example, while not giving it to mine."

In other words, I began thinking to myself as I stepped outside and, following Dr. Green's directions, more or less moped my way from Lab Building No. 9 over to the NIH Clinical Center, I shouldn't think of a particular God as being the author of my fate, but rather of his or her secretary, some grossly overworked and underpaid scrivener who just happened to make an innocent typo on page 2,500 of my own personal thirteen-volume set of the so-called human encyclopedia.

From a red-canopied snack wagon in the Clinical Center's front lobby, I bought a can of Coca-Cola for my hangover, and then waited out the last half hour until my appointment, seated on a wood bench among a set of tall glass display cases that held an exhibit dedicated to a Dr. Marshall Nirenberg and the Nobel Prize–winning DNA research he had apparently conducted at NIH back in the 1960s.

The cases contained a number of hulking, decidedly dated-looking machines not unlike the lathes, punch presses, and drilling machines that I remembered from the tool-and-die shop at my father's National Tel-Tronics plant, anomalously cumbersome equipment, it seemed to me, for the sort of work described by the little cards in the foreground of each of the display cases.

"These are the machines," explained one, "that Dr. Nirenberg used to break open the cell walls of a bacterium and from the escaping cell sap decipher all the working components of protein synthesis in our genes." The very Chinese boxes, in other words, that Dr. Green had just been describing to me—"bacterial DNA," I could hear him explaining to me at that very moment, "because it's exactly the same stuff as human, which is the same as that of yeast or frogs or cows or plants."

Across from the display cases was a set of glass doors that opened back onto the grounds of the NIH campus. People were sitting out in the grass under trees, sipping coffee and chatting. Someone was tossing leftover lunch bits at birds. There was the whole tableau before me: the living world, a varied assortment of the same common DNA clay, and yet only our arrangement of it having gotten the few last tweaks and taps—as though from the hands of a late-night, punchdrunk sculptor—that allow us to regard our own and all the other arrangements.

Not only to wonder what it would feel like to be alive, at this very instant, inside the human being just beside you. Whether you would topple over dizzy and nauseous from the difference, or pass as easefully into their body and mind as you would into a warm, outsized sweater?

But to wonder, as well, what it would be like to be, right now, behind the eyes of that scrounging sparrow, or some head-swiveling pond swan? What it would be like to be one of those DNA arrangements that live blissfully unaware even of their own fatal inner misspellings: flowers with ill-shaped petals; a stunted, misshapen tree limb—and perched, perhaps, somewhere out along it, a whole family of cold-fluttering, flightless flies.

Chapter
10

D R. FANANAPAZIR WAS ON THE PHONE WHEN I appeared at his office door. He motioned for me to take the seat opposite his desk.

"Yes," he was saying. "I see. I'm sorry. When exactly . . ."

Medical files were stacked on the floor all around me. On the wall just above my chair was a huge map of the United States. Pins with tiny white pennants were stuck from coast to coast in clusters of varying density. Surrounding the map, and spilling over to the office's side walls, were hundreds of photographs, of patients, I surmised, those enlisted in the HCM research protocol. The majority were children, infants, toddlers, and grade-school kids, smiling before those deep blue, neon-streaked backdrops of cheap, department-store portraits. There was one of a teenage girl, from what appeared to be her modeling portfolio. She was wearing a red turtleneck sweater and a short white dress, her left leg propped up high on the chair before her.

"Happily entered heaven, October 1994," were the words written beneath the picture. "Thank you, Dr. Fananapazir, for all your efforts."

"Yes," he continued, head down, phone scrunched up between his right shoulder and ear as he wrote notes on a legal pad. "Well, please keep me informed."

"A patient of mine," he said, hanging up the phone, his eyes bypassing me for one of the photos on the wall over my shoulder. "A young man. Forty-five. His heart went into a near-fatal arrhythmia just this morning. If he recovers, we'll bring him in for further examination."

I turned in the general direction of where the doctor was looking, and then my eye was caught by a different, nearer close-up, that of a wide-eyed infant with a full head of blond curls and the faintest rim of blue across his forehead.

"We couldn't figure out what that was at first," explained Dr. Fananapazir when I asked about it, a deeply distracted man with paunch-loosened shirttails. "Then we realized that the child's heart was going into very slow rhythms, sometimes stopping completely. He kept passing out and bumping his head on things."

The doctor sat in silence, waiting for me to turn back toward him, the very tactic he'd employed earlier on the phone. I promptly stood up and handed him my father's medical records. He began flipping through them. I walked back over to peruse the wall of photos, playing a pointless game of trying to guess which faces matched which of the map's pennant pins.

Here was a wall map that was in and of itself the result of centuries of exploration, conquest, disease, and death, and yet now it seemed to have been superseded, rendered nearly incidental by the current new-world map of our genes. It is one that, like the earliest cartographers' renditions of our world, still has its own vast stretches of dragon-filled *terra incognita*, and yet a map by whose otherworldly reckonings I was about to be declared the next-door neighbor of complete strangers because of some potentially bad fly-wing muscle fibers sputtering away inside our hearts.

I returned to my seat, watched the doctor flip through

the photocopies of my father's medical files. I'd already looked through them on the train ride out to Bethesda that day: the records of my father's hospital stays and doctor's office visits, furled, dark-edged, drearily dull documents, which, nevertheless, now read to me like prophecy. The pages all followed the same pattern. First came a physician's barely legible written report, followed by the accounts of the various tests and medications administered, my father's ever-graceful signature at the bottom along the line labeled PATIENT'S CONSENT. Then came the electrocardiogram or EKG readouts, page after page of little jagged peaks, flat plains, and sudden deep valleys: the precipitous topography of my father's heart.

Dr. Fananapazir seemed to dwell the longest on the top set of pages, the hospital report of that very first episode on that summer morning back in 1967 when I was away in the Adirondack Mountains: "Siebert, Charles J.," the top of the first admission page reads, "III Orchard Rd., Ossining, Date of Admission 8/12/67 11:30 A.M. Born: 4/13/20, Catholic, White/Male, Age 47."

The rest of the page is in the attending physician's handwriting, an abbreviation-studded scrawl. Under the heading "Chief complaint" there are only two words: "Chest pain." Under "Present illness," the first paragraph begins: "47 y.o. who has been essentially well until today . . ." and then came sentences that would bring me upright in my subway seat, confirming, as they did, my long-held suspicion that my father's heart had begun to go awry years earlier without his letting on about it: "Pt.

[patient] had an episode of transient chest tightness and was SOB [short of breath] in 1964. This cleared spontaneously, lasting only a few minutes. In 1965, had another such episode. Seen in my office a day later and EKG was normal. Complete P.E. [physical exam] then was essentially normal."

The next paragraph I found striking for its sudden shift in tone, from the coldly clinical to the breezily prosaic: "Today Pt. was having his morning coffee when he suddenly developed squeezing, severe chest pain radiating into both arms and into his neck. He was SOB and became dizzy and cold."

And high up on a mountain. And deep within its cold, dank earth, you pass, quietly, alone, the panicked cries of the others around you only driving you further into your self and forward, the cave walls beginning to squeeze in and the water to rise up around you, exactly as it does inside of us when our hearts begin to fail.

Doctor Fananapazir closed the files, folded his hands in front of his lips, and for the first time devoted his full attention to me.

"There is," he began, "as I've said, a fifty-percent chance that you have inherited this condition from your father."

Again, he went silent. I briefly considered pointing out to him the little checks that I'd made on the subway next to a series of seemingly insignificant entries in one of the nurse's logs of my father's first hospital stay. Each time a family visit is noted, the subsequent entry reads: "Family's gone now. Pt. looks weak and exhausted."

This, it seemed to me, strongly supported the too-many-children theory for his heart's failure, even while it put me in the unprecedented and somewhat craven position of claiming to be an accessory to my own father's death. I decided, however, to hold my tongue and, this time, successfully waited the doctor out.

"Of course," he finally offered, having no doubt noticed the utter lack of blood in my face, "there is a fifty-percent chance that you have not."

Bedside manner, it occurred to me, has yet to be cultivated in the genetics ward, perhaps because there are no beds there, no furnishings of any kind, just blanched, borderless rooms filled with pressed gel plates and gently humming scanning computers, the meta-scriveners, the new faceless goddesses of fate.

I sat there listening to Dr. Fananapazir talk about the various genes linked to HCM—one found on the "short arm of chromosome 1," one each on "the long arms of 14 and 15," and a new one they've found someplace along chromosome 11, "we aren't exactly sure where"—and as he spoke, the map of the United States on the wall above me began to blur into oblivion.

I was being directed now to strange new localities in some as-yet-uncharted world, a world that—perhaps because I was so haunted as a child by the new Chilmark development we'd moved to from Brooklyn back in the early 1960s—I could only imagine as one of those very developments, at night, all those white pennant-pins now arranged like mailboxes along endless, winding rows of neatly arrayed driveways: the seemingly quiescent but inwardly roiling cul-de-sacs of our chromosomes and genes.

"Each gene related to HCM can have as many as thirty different disease-causing mutations," Dr. Fananapazir explained to me, "and the risk varies from one mutation to another. In some families, the symptoms develop at a much later age and the disease has a benign diagnosis. In other families"—he glanced furtively at one of the file boxes on the floor beside my chair—"they have a mutation that wipes out everyone before the age of fifty."

At one point, Dr. Fananapazir went to a file drawer behind his desk, then came back and handed me a copy of an article he had written. It describes the variations among the effects of the specific mutations on the most prominent gene associated with HCM, something called the "beta-myosin heavy-chain gene." I couldn't make much sense of it, but one detail did catch my attention: the way that each mutation was identified as a different number, the "403" or the "908."

"That means," Dr. Fananapazir said, "the mutation occurs at the 403rd codon, or the 908th codon within the beta-myosin gene."

Codons, I remembered from Dr. Green's impromptu lecture, were those little amino-acid-wielding protein builders. They, I now understood, were the ones who lived in those different houses I'd imagined set out along the wilted, misshapen X's of our chromosomes, each with their numbered mailboxes, the addresses, in this instance, of the houses of heartbreak. And then I was shown into one such residence, at codon number 403, within the beta-myosin gene on the long arm of chromosome 14, where I learned just how surreally specific and arbitrary heartbreak can be.

The myosin gene, as Fananapazir explained it, "codes for myosin," one of the main proteins of the heart. Myosin is what's known as a molecular motor. It works in concert with another muscle protein, called actin, to coerce and control the contraction of muscle fibers—the ones nature invented to power fly-wings and hearts.

"Here, I can show you," Dr. Fananapazir suddenly announced, escorting me out of his office through a series of corridors that led into the heart of the Clinical Center's research department. We came to the open door of a laboratory, one in a long hallway of labs, each one a mirror image of the preceding one and—like all modern biology labs—not at all revealing or reflective of the kind of work being conducted there.

Dissection trays, scalpels, formaldehyde, actual tissue, flesh, blood, these are the trappings of the old, the exhausted, the fully found world. Now the lab workers themselves seem outmoded castoffs, like those cell-splitting machines I'd been staring at in the downstairs lobby: overly plodding, ham-fisted amalgams of the very minutiae they seek to unravel, trying, with their thin glass tubes, gel plates, pin-headed droppers and color dyes, to briefly stay and study the wriggling wraiths of our inner essence.

Dr. Fananapazir stepped inside the lab door, knocking as he passed. A tall, slender, dark-haired figure emerged from the back of the lab, making his way past a number of white-coated assistants. He was wearing glasses remarkably similar to the ones I'd recently uncovered in my father's box.

"Dr. Neal Epstein," said Dr. Fananapazir by way of introduction, Neal and I shaking hands as we all moved toward his cramped back office. There, upon Dr. Fananapazir's request, Neal called up on his computer an actual image of actin filaments being pulled by myosin: microscopic glowworms sliding past one another, like tiny hands, pulling, one hand over the next, on a chain. I stood there kind of leaning into myself, feeling those flies alight again now on my heart, billions of squiggly protein heads coaxing it into its very next beat.

Dr. Fananapazir thanked Neal for his time and then ushered me back to his office. I was next shown what he described as a first ever 3-D X-ray crystallography photo of myosin. It had a round head and a spermlike tail. Myosin, he explained, his voice now registering at a slight, dreamlike remove that I hoped signaled an imminent conclusion to this day's marathon science lesson, is composed of thousands of amino acids, the precise order of which is arranged by those codons in the myosin gene in the course of protein synthesis.

"But people who have a misspelling at codon 403," he said, "get one wrong amino acid in that chain of thousands."

This is what a deadly mutation is. Each of our genomic texts is filled with such minuscule misalignments. God's secretary's typos. Most never manifest themselves.

Others may cause an allergy or a slight asymmetry between our left and right nostrils. But a tiny typo at mailbox number 403, up at the top of the long arm of chromosome 14, kills you. It misshapes the head of myosin just enough to impair its motion, causing the heart, second by second, each day, over the course of a life, to contract improperly and thus thicken and fail.

Chapter

11

S ECOND OPINIONS, IT SEEMED, WERE ALSO NOT A feature of the genetics ward. Dr. Fananapazir immediately began discussing with me the details of the research protocol, going so far as to suggest, as he got up and showed me to the door, that I make an appointment with his secretary to begin the series of preliminary tests required of all protocol participants.

I'd been pulled by then so far down into the particulate underpinnings of our existence—that incalculable mix of intracellular bases and acids which, in all their fulminations, somehow give rise to that thin, vaporous film we call consciousness—I felt I'd never be able to make my way back up to see and enjoy again that film's overall, pervasive spell.

"Is tomorrow okay?" I heard Dr. Fananapazir's secretary asking. "Three P.M.?"

"Yes," I said, "fine," and then started away. I got nearly to that point where it would have been too embarrassing to turn back, and then I did.

"Could you," I asked, "tell me again the way to Dr. Epstein's laboratory?"

The place looked to be shut down when I arrived. Neal's assistants had gone for the day. All the lights were off but for one slanting slab of pale yellow broken across a back lab table. I walked to the door of Neal's office and peered in. He was exactly where Dr. Fananapazir and I had left him, sitting at his computer, jotting down notes on a legal pad as the same image of myosin and actin still danced on the screen above him.

"Nature was confronted with this problem," he said,

barely acknowledging my presence in the doorway, the intensity of his focus drawing me quietly inward, like a student arriving late to a lecture. (I'd learn only later that Dr. Fananapazir's secretary had called ahead to warn him of my arrival.) "The problem of how to generate enough power to move an insect's wings 150 times a second using the molecular fibers that were available.

"Nature has motors and it has generators," he continued, his voice still thick with the telltale vowel distortions of a suburban Boston childhood. "The generators that supply the cell's energy, for example, are the mitochondria. They provide energy in the form of a three-phosphate molecule called ATP, which has intrinsic energy that can be released by cutting off a phosphate. That energy is then transferred to the bonds in the molecule myosin, winding up the spring of the myosin motor, which then grabs on to and moves this actin filament and that's how these two filaments can move like this, can slide past each other. That's what muscle is, a series of these little contracting elements that makes the larger whole contract."

"ATP?"

"ATP! Adenosine triphosphate!" Neal removed his glasses, eyeing me wearily. "It's a chemical fuel."

This has since become a regular dynamic between Neal and me: he excitedly launches into the abstruse particulars of his research and then has to double back to rescue me from the shores of stupefaction. It is something he is uniquely adept at doing, having begun his professional life as a Ph.D. candidate and teaching fellow in philosophy at the University of Wisconsin. In 1976 he made the unlikely leap from Kierkegard to cardiology. Our equally unlikely friendship, I've always suspected, is fueled in large part by Neal's ongoing delight over the unexpected ways in which my ignorance forces him to recognize and reconfigure with his old philosophical overview the terms of his new and incredibly recondite knowledge.

"ATP powers most everything in our bodies," he said, just shy of shrieking. "You've got to have energy to drive things, to maintain the dynamic force of life.

"Life," he paused, revving up now to draw one of his signature big pictures, "is kind of a matter of anti-disorder, of anti-entropy. Things work because concentration gradients are met, like you have more sodium on the outside of one tissue wall and more potassium on the inside and these things have to be kept in some kind of dynamic tension. To do this you need to have motors, generators and pumps, pushing this out, and that in, and taking this there. This is the dynamic force of life, this tension between the way things would go if they were just left to diffuse, and where they are kept as a result of energy. Life is an aberration, an exception, because all matter naturally winds down toward entropy. But for a brief time you have things the way they wouldn't be if they didn't have energy keeping them that way, energy maintaining these brief pockets of anti-entropy that is a life."

I asked Neal at one point whose heart that image of myosin and actin was taken from. He told me it was a cross-section of heart muscle belonging to a "transgenic mouse," one of many mice to which Neal had purposefully given a faulty human fly-wing muscle fiber gene. He refers to them as "Flice."

"Nature had already devised these incredibly efficient insect flight muscle fibers," he continued. Nature in Neal's ethos is always spoken of as the ultimate artificer, much like the head tool-and-die man in my father's company machine shop, picking and choosing from among the available preexisting molds, and sometimes having to forge entirely new ones, all by way of shaping the required parts for the next job.

"And when it was confronted with the problem of designing something like our heart," he continued, "a muscle that has to beat some three billion times in an average seventy-year lifespan, it would clearly need any trick in the book to do this, and ultimately it used one of the very designs it had devised for the insect flight muscle."

Neal must have noted at some point a queer look of exhaustion and quiet terror in my eyes. He got up from his desk, fixed us both a cup of coffee from the little four-cup machine he had wedged among stacks of papers on his office windowsill, and sat back down. My mind riddled by now with images of speeding fly-wings and squiggly protein heads, I eyed my coffee a bit warily, wishing I'd had a stiff bourbon instead.

Neal now began asking me very basic questions about my family history; my health and that of my brothers and sisters; what I knew of my father's heart condition, and of the others on both my mother's and father's side of the family. As I spoke, he ripped off the top page of the white legal pad he'd been writing on, and began to draw what would become my family tree.

He started with a central circle and square, signifying my mother and father respectively. He shaded in the square to mark the possible presence of the HCM gene in my father, and then crossed a line through the square to indicate "deceased." He then asked me to start calling out the names not only of my brothers and sisters and their children, but of all the relatives I could think of on both my mother's and father's side.

This would make, in the end, for a sadly misshapen growth, my father's few known German relatives, all deceased, dangling like some long-withered winterberries against a vast white icy tundra. On the opposite side of the page, meanwhile, were the ripe, multibundled branches of my mother's Sicilian heritage.

Neal listened and drew. When the tree was complete, he then began to call the names back to me. He said that just hearing them aloud often helps people to remember details about their families. He'd call the names out, one at a time, these simple sounds I've known my whole life, and yet their very invocation somehow instantly recomposing in my mind the people being named, and they, in turn, helping to compose me, lift me up whole again out of the dizzying particulateness of my own paranoia.

I felt like one of those figures in a Chagall painting, adrift above his village's rooftops, above, in this instance, the most familiar landscape of all, my mind calling into view, with curious clarity and calm, not just the members of my immediate family, or my numerous Italian relatives, but even those few little-remembered or never-known figures on my father's side, the side I'd long been content to let remain in the shadows, certain that the fateful fly-wing curse must, if it existed at all, surely lurk there.

I recalled at one point a picture I'd seen once of my paternal grandmother, Madeline, maiden name Rettig, whose early death and obscure family history rendered her a complete dead-end in the potential search for the telltale genetic typo. Taken in 1919, when she was twenty-three, a year before my father's birth, the photo reveals a petite, thin-wasted woman with pale, rounded cheeks and a broad nose, her brown hair drawn back in a tight bun behind her head. Dressed in a long black skirt and high leather lace-up boots, her V-necked white blouse frames a wooden cross on a necklace, the cross resting just above her breastbone.

Her left arm is draped around the corner post of a wood fence bordering some tiny Lower East Side tene-ment yard, while her right arm is being clutched by my grandfather, Charles the First, twenty-seven years old, just home from the war in Europe, where he'd received the Cross-de-Guare from the French for bravery in action. He's still in his cavalry brigade uniform, his tasseled rid-ing hat held down at his side, the two of them smiling broadly beneath the scrawny, sun-bleached branches of a young ailanthus tree.

Charles the First, the family's first purveyor of anonymous parts as a longtime employee of the Arrow Supply &
Tool Company, was sixty-four when his heart gave way. His
coat pockets were found filled with crumbled crackers
from our duck-feeding expeditions. Still, as Neal would
remind me, it isn't known whether his heart's ultimate failure was even the result of HCM, to say nothing of whether
any other genetic typos might have been involved.

My grandfather had three older brothers. The oldest,
Martin, was born in 1877, according to the New York City
census report I obtained for the year 1900, and worked as a
meat packer, a noteworthy fact to me only because of my
own brief history in the meat industry. But the possibility
of a meat-working gene in the Siebert blood notwithstanding, I could find nothing about the condition of
Martin's heart or the manner of his death.

John, born one year after Martin, was my father's
favorite uncle and surrogate father, thanks to Charles the
First's constant sales trips for Arrow Supply & Tool. John,
a bookbinder by trade and the only one of that crop of
dour Germans ever seen smiling in photographs, was
found at age sixty-seven sprawled on the tile floor of the
upstairs bathroom of the house on East 37th Street, also
the victim of a heart attack, but again, of unknown cause.

And then there was Henry, a year older than Charles, a
clothes salesman. He doesn't appear in any of the family
photographs I've ever seen, and what I knew of his life
didn't offer up any clues about bad genetic inheritance

insofar as Henry eclipsed the natural arc of his own heart's flight at age 52 by drowning it in alcohol.

"There are just so many variables," I could hear Neal saying in the background, my mind alighting now even on the most distant known figure in the shadowy Siebert lineage, my great-grandfather John Siebert Sr., born in 1853. A New York City census report from 1880 has John living on East 6th Street, prior to the move to the house on East 15th where my grandfather and father were born. John was a chandelier worker most of his life, until, family legend has it, one day in 1902, when, having apparently stared too long into the lights of the Waldorf Astoria's entranceway fixture, he flailed dizzily out of the straps of his support harness and came crashing down to the lobby floor.

"Why did your father's heart problems set in so much sooner than his father's," Neal asked, "or than his uncle's did? What effect did your father's job and lifestyle have? How did intangible factors like stress and daily temperament affect his genetic predisposition..."

As he spoke, my mind began to course again over the contents of that box: those cursed eyeglasses; the Exec-Mate diaries with their dutiful directions; the various diplomas demarcating his undeveloped promise; the report cards of all those proclivities unpursued—the lot of it beginning to wriggle awake with the sound of Neal's voice, bursting open the box's cardboard lid, and taking flight around my hotel room.

Chapter

12

JUST WEEKS AFTER THAT FIRST MEETING WITH NEAL, I found myself driving alone on a highway toward the heart of America, toward one of the towns I'd seen marked on Dr. Fananapazir's wall map.

There are, on the genome map, any number of biologically cursed places, like the three Venezuelan villages on the shores of Lake Maracaibo where hundreds of people from the same multigenerational family have the mutation on chromosome 4 that causes Huntington's disease, a fatal, degenerative brain disorder marked by uncontrollable jerking and writhing throughout the body, suicidal depression, and intellectual deterioration.

And then there is the tiny Midwestern town I arrived in that summer on the eve of its Fourth of July celebration, a town two traffic lights long with a flag-bedecked granite courthouse in the main square and a rebuilt homesteader's log cabin on the lawn beside it. A town I can't name because among the many prospectors now working the new, uncharted terrain of our genomes are insurance company investigators who immediately delete from their rolls people who have defects in their genes like the HCM mutation on chromosome 14 that looms over one large multigenerational family in what I'll call Lawton.

I stopped at Lawton's main square. People were sitting on sidewalk benches in front of the courthouse, watching through the late-slanting sunlight the town's fire chief open the weekend festivities with a fire-extinguisher demonstration. I asked no one in particular if I was heading in the right direction to the house of John and Karen Alston, one of those ill-fated families in Dr. Fananapazir's file boxes. Lawton being a town of about 2,000 people, everyone knew the way, pointing me farther down the main road.

It turned out to be only a couple of miles more through cornfields, but it seemed to me an interminably long stretch. Neal had phoned the Lawtons on my behalf, explaining both about the book that I was intending to write and about my father's own history with HCM. They readily agreed to meet with me—were, as Neal assured me they would be, happy to do it. And yet, as with the heart harvest itself later that winter, the closer I got to the very encounter I'd gone out of my way to orchestrate, the more uncertain I became about my motives for wanting to.

Their house was a one-story, trailerlike structure with log-cabin siding set at the edge of a small patch of woods surrounded by cornfields. I was met at the door by Karen Alston, a very pretty, wholesome-looking woman in her early thirties. Her three young children were pressed in around her legs, peeking up at me. Her husband, John, the bearer of the bad gene, was working the late shift at a factory a few towns over.

"I married into the curse," Karen said, smiling broadly as she escorted me to a plaid sofa in her crowded living room.

She told me that Karla, nine, their oldest child, does not have the gene. Mary, seven, does. She began to develop symptoms a year after the birth of her younger brother, Arthur, the blond baby I'd seen on the wall of Dr. Fananapazir's office, with the rim of blue on his forehead. Arthur, five now, had a pacemaker implanted when he was still a baby. It had not only righted the rhythm of his heart, but, to both Dr. Fananapazir's and Neal's surprise, had also helped to reverse the thickening of his heart's muscle walls. Arthur was nonstop action while I was there, wrestling with his sisters, swinging mini baseball bats, eventually getting chased out of the house by his mother.

Karen explained that it was John's father, George Alston, who had passed the HCM gene on to all four of his children—Barbara, Sophie, John, and John's younger brother, Jim. Since learning that he was the carrier, George has made it a custom to buy all of his grandchildren life insurance policies on the day of their birth.

"He feels bad about the gene," she said. "It's kind of his way of making it up to everyone."

Life insurance, it turns out, is the only type of coverage that George, who can't get insurance for himself, can secure for the children, provided, of course, that they are born healthy. Arthur, however, emerged from the womb struggling for breath, and Karen would have to administer CPR to him four times in his first eleven months of life. He still has no bottom front teeth, all of them lost as a result of oxygen deprivation, his brain having grabbed every last ounce during his repeated blackouts.

"He'd get this look," Karen said. "He'd be running around and then just stop. His eyes would get all glassy and fixed, and he'd turn a bluish gray. Sometimes I'd have to give him a shake, kind of jump-start him again. I never let him out of my sight."

When Karen's daughter Mary first started having her chest pains and fainting spells, Karen tried to keep her close by as well. Most nights she would sleep holding baby Arthur to her chest and her other hand pressed to Mary's. It got to the point where the slightest bad rhythm in their hearts would wake her.

"There are times," she said, her voice cracking, "when I get so angry at the gene, at John, at God, wondering why he chose my children. Because he thinks we can handle it? Is this some sort of test? John blames himself, and I know John's dad does. I tell them it's not their fault."

Karen mentioned more than once that she and John would not have had children if they'd known then what they know now. They scratched plans to have more children when Arthur came along, and have had to to put all other plans on hold because of the tenuousness of their children's condition. Much as he'd like to, John can't leave his factory job for a higher-paying position elsewhere, because his company's insurance plan, which automatically covers any child born while he's employed there, is the only way he can get coverage for his family. They've tried several times with other insurance companies, but none will touch them.

"We had it all planned out," Karen said. "Have kids and then save enough money to build the home we want. Now it's one day at a time because we just don't know what's going to happen."

I asked Karen if she and John had spoken about the disease before they were married, before the genetic connection with John's HCM had even been confirmed.

"Oh, yeah," she said. "He even told me maybe I should marry someone else and have kids because there's this disease in his family and he might pass it on. I'd say things like 'You're okay, and besides, they'll have a cure by the time we have kids.' Guess I was a bit blind. You are when you're younger."

Alone in my Lawton motel room that night, staring into the bathroom mirror, I wanted to tear my father's features away, hating the very hands that I would do it with for having so much the look and the lay of his. A geneticist would tell me that I was now confusing surface phenomena with my inner, predetermined genetic fate, genes having become in our minds the sole arbiters of that now.

Somehow, the closer we've gotten to a true scientific understanding of them, the more primitive our conception of genes has become. They've achieved a kind of deification, are seen as the pure, unbiased prime movers not only of our ultimate fate but of all we are and do up until our deaths.

Genes are thought to be responsible now for everything from poverty to privilege, from misdemeanors to murder. Criminal lawyers have already begun offering as a defense the presence of a "criminal gene," which they claim runs in a client's family. It's as though we're looking to biomolecularly deconstruct that richly textured and inherently ambiguous text that our own biology has allowed us to revise endlessly. The scales of the ongoing nature-versus-nurture debate are so tipped that it is perhaps more pertinent to ask, Where do genes finally leave off and people begin?

At what point, I began to ask myself that night in Lawton, do we stop letting this new knowledge about the influence of genes upon our lives influence our life's choices? At what point do we stop letting, to use geneticists' jargon, genotype (the people we're basically blueprinted to be) override all those other considerations that govern phenotype (the very singular people that we each, with our own unique set of experiences and responses to those experiences, develop into).

Had my father gone for a test and learned about his alleged HCM gene, would he have chosen not to have had the family of seven that has thus far, and for reasons no geneticist can explain, emerged completely unscathed? And even if some of us start to develop symptoms in the coming years, haven't we already had substantive lives, relatively pain-free, good lives compared with what some have had to endure?

And what if I were to call up Neal right then and tell him I'd like to be part of the protocol, and it eventually revealed one of the culpable mutations? Where would that leave me? A fully reconstituted and justifiable hypochondriac, waiting around—fingers pressed to pulse—for the other shoe to drop.

I might, too, have a benign mutation, one that is, as Dr. Fananapazir phrased it, "compatible with a long life." Or I might be what's known as a "skip," one who has the mutation but never develops the symptoms and instead passes the bad gene on to the next generation. I might also walk outside tomorrow and, as happened to my uncle years ago, get killed by a falling piece of building eave. A darkness awaits us all. What do I need with a slightly refined set of odds on the advent of that darkness?

I thought back to Neal sitting in his office a few weeks earlier, drawing the family tree as I called up the names and as many details as I could about the people behind them, raising, as I did so, more and more questions that genes alone could not answer. Questions concerning all the inconsistencies that can be present within the same family known to have the mutation. Why, for example, were John and Karen Alston's children so hard hit and the other Alstons not? Could their spouses' genes be mitigating the bad ones in unknown ways? And why, if both John and his younger brother Jim have the gene, has only John struggled with it all his life, suffering blackouts, chest pains, and low endurance?

No one could tell me why my father's heart problems had set in so much earlier than his father's or his uncle's, or what effect a relatively inactive lifestyle and a disappointed disposition might have had on his genetic predisposition, those immeasurable aspects of any heart's story: stress, submission, faintheartedness. Little is known about how the daily movements and encounters of the very body and mind that the genes help to create, effect, in turn, the ongoing expression of those genes. Most diseases, in fact, do not have a direct, one-to-one causal relationship with a specific gene. They are, instead, the result of the same unfathomably subtle interplay of gene, gene expression, and environment that has formed each one of us, an intricate biological fugue of which we've been able to identify only a few isolated notes.

"The thing about a human being," I remember Neal saying to me, "is that if you take apart all of the proteins we're made up of and throw them together again in a test tube, they'll just sit there. You can take one cell and put a gene in it and then from it raise up a bunch of daughter cells, all of them with the exact same genetic makeup, and within each daughter cell, there will be a different expression of that gene. That's just one cell. Now imagine all the cells in an organism as complex as a human being is, and how they develop in time and space and in response to other developing parts of the organism—and all these amazing numbers of intersecting events have to occur in just the right amounts and sequences—and you wonder how there can be life at all. When it does occur, you get variability. You think your right eye is the same as your left? Now think of children. Two parents contribute half of their genome to their child, and they end up basically with a stranger."

The Alstons had invited me to their annual Independence Day picnic the following afternoon. On my way there, I stopped by Lawton's main square to watch the town's holiday parade. It took all of ten minutes, a brief pass in front of the courthouse and then a right turn out of town: police cars and fire engines, an ambulance, a mixed group of war veterans—World War II, Korea, Vietnam—two circus clowns on mopeds, a bevy of area high school beauty queens seated on hay bales, a procession of tractors, and, finally, the "Citizen of the Year" in a yellow convertible.

On the way back to my car, just off the main square, I passed a train depot that also doubles as Lawton's historical museum. There was no one around, but the door was open, so I let myself in. Among the exhibits—kitchen utensils and farm equipment from the nineteenth century; a giant loom beside an array of weavings by local village women; the uniforms of Civil War officers—there was one devoted to early medicine. Neatly arranged in a glass display cabinet were old lancets and forceps, an assortment of needles, a medical saddlebag for house calls on horseback.

In a separate display case was a phrenological bust, used in the once widely regarded nineteenth-century practice of trying to measure a person's character and mental capacity by feeling and diagramming the contours of his or her skull: an earlier, much cruder attempt at mapping the human essence. As I stood there surveying the different puzzle-piece quadrants: "Individuality," "Animal Propensities," "Selfish Sentiments," "Conservativeness," "Destructiveness," and so on—it occurred to me that for all the dizzying precision and exactitude of current genetic mapping, we're really not that much closer either to pinpointing the locales or the amounts within us of the very characteristics labeled there before me.

Though genetics will likely enable the development of countless cures and therapies for disease, what ultimately confounds genetically determined explanations of us is the very fact that our DNA assemblage—whether by accident or owing to some "divine" force—is the one that spilled over into consciousness and the singular capacity (or curse) to regard and argue with that assemblage.

All that human beings have created, all of culture and its history, is, in essence, the result and the record of that argument with our biology, one that, in turn, continually redounds upon and influences the shape and expression of that biology. It is as constant and variable an exchange as that ongoing daily one between our heart and brain. You and I are, in the end, at once as common and unknowable as the makeup of our very next thought, or the direction of the walk we'll take later, and what happens in the course of it, and how it affects us.

I came upon Karen and John Alston standing in a huge clearing amid the ongoing rows of corn behind John's father George's house. The other members of the Alston clan were playing softball with some folks from town. There are hundreds of Alstons scattered around the country, fifty or so in the Lawton area alone, the rest in Michigan, Florida, and California. A few years ago, Neal and Dr. Fananapazir managed to gather the lot of them in Chicago to conduct one of their pedigree studies.

"Everyone was pretty cooperative," John told me in his sleepy drawl as he watched little Arthur smack a line drive and then tear off in the general direction of first base. "There was one cousin, though, who finally got fed up with all the talk about this killer heart gene and punched a hole in the wall right above Dr. Fananapazir's ear."

It did make me a bit uneasy, standing there, watching so many suspect hearts dashing around the bases, but none of them were holding back. I found John's father leaning against a pasture fence, beaming at his grandchildren, cheering them on. He still has, he told me at one point, the regular blackouts, the chest pains, the shortness of breath, and yet, at age fifty-seven, he continues to work his daily nine-to-five shift at Lawton's local iron foundry.

"What can you do?" he said, a big smile on his ashen face. "Maybe the folks at NIH will come up with something."

We ate fried chicken brought out in buckets by the owner of a local restaurant. I talked with George's daughters—Barbara, a mother of two, and Sophie, of four. Both said they'd go ahead and have their children again, knowing what they do. Sophie said she can't even bring herself to find out the results of the tests that had already been done at the NIH on herself and her children. Barbara and one of her sons, Seth, know they have the gene; Kerry, the twin of her son Carson, who is curse-free, was born with a malformed heart and died when he was three days old.

"Yes," Barbara said, "I'd have my kids again. I live for my kids."

When darkness fell, Jim set off fireworks above a man-made pond dug behind the house. Later, as I was saying goodbye, George took me aside to say that I or anyone in my family should always feel free to call them to talk about "this disease."

As I was heading off to my car, I turned to see Jim following behind. He wanted to explain that if I was going to write anything about my trip, he was the reason I couldn't use his family's real name. He said that everyone else had already lost their various insurance battles, but he and his wife had just made the decision to have their own children, and Jim was still holding out hope for them.

"Do us right," he said.

I drove away, a clear dome of stars capping the dark rows of corn all around me. I'd pretty much decided by then that there would be no genetic test in my life, that I didn't need to know. I would drive all night to get back home—to get back, in a sense, to my own unknown. I had a right to that, at least. Not simply for the privilege of hope but of pessimism, my own rendition—however mis- or self-guided of how and when I might die. That, in tandem with the necessary daily dismissal of our own mortality, the moment-by-moment illusion of our ongoingness, is what frames that story by which we get from day to day.

I remember at one point during that first visit with Neal a few weeks earlier in his cramped NIH laboratory office—the image of myosin and actin dancing on the computer screen behind him; my mind already deeply attuned by then to our fellowship with flies, and all the other DNA arrangements—it suddenly occurred to me to ask him if there was such a thing as a creature on this planet that didn't have a heart of some kind or another.

Neal paused for some time, struck, I suppose, by the blunt-edged simplicity of the question.

"Well," he began, "how about any single-cell organism?

You don't need a heart in a single-cell organism. If you have one cell, it does everything. An amoeba doesn't need a heart because all of the protoplasm of life is surrounded by the membrane that defines that amoeba. It is surrounded by its food and anything it needs it can get, because its entire perimeter is accessible to the environment that it feeds off."

"But," he continued, finger raised, signaling the setup for another one of his big pictures, "when you start putting cells together, when you have a multicellular organism, you have different kinds of cells, and these cells, as they develop and get more specialized and better at what they do, like becoming a brain, for instance, or a heart, they lose the ability to do other things, like acquiring their food and disposing of their waste.

"These are cells that are not going to make it on their own. They'd be poisoned by the waste they produce in the very process of being. They couldn't make it without the bathing and nurturing environment created by the heart. When, way back in biological time, nature started putting more than one cell together, it had to build a heart. It's the complexity of the multicellular organism that first impelled the creation of the heart."

Neal and I would end up down in the NIH cafeteria that evening, eating dinner, talking all the while about fly wings and phosphates; about "Of Flice and Men" (a title Neal playfully considered giving to one of his papers on the insect flight response); about nature's "borrowed tricks" and "repeated designs" over the course of evolution, and the compelling "homologies" to which such inspired borrowing has given rise, like that between fly-wings and our own hearts; or that between the increasing specialization that the heart has allowed our own cells to develop, and the very same increasing specialization of our knowledge and academic disciplines, which the bathing and nurturing medium of civilization has, in turn, allowed us.

I began to take, even then, a deep sense of comfort from such knowledge, the inherent mystique of modern scientific minutiae freeing me at once from the limits of my paranoia and the dimly wielded dogma of theurgic conceptions of our origins. It was as if I were being introduced to a whole other modern-day creation myth, a new mystery of faith, one arising out of the very scientific specificities which—it has for far too long been our conceit—are supposed to dispel myth and wonder and faith.

It makes nearly too perfect sense, I remember thinking to myself in the course of the long drive home from Lawton that night: the fact that the heart was invented in order that life might complicate itself. The heart is not the seat of the soul, of the intellect, or of the emotions. It is the soul of all of those, the vital, animating spirit of the very complexities that will one day wear that heart down.

Our hearts, I felt certain no one flying by me on the highway that summer night would have been surprised to hear, were born to be broken.

Chapter

13

THE PARKING LOT OPPOSITE COLUMBIA-PRESBYTERIAN'S Milstein Hospital building was nearly empty when I pulled in. I sped to a space directly alongside the ticket booth, told the attendant that I didn't know how long I'd be, but certainly overnight, then raced across the lot, striding over rain puddles, my cross-trainers pressed into earlier action than I'd anticipated.

I got to the hospital building entrance a mere six minutes behind schedule. There was no sign of Dr. Rosen. Thinking that he might have gone ahead to change, I was about to run to the front desk just inside the lobby entrance to call his beeper when he appeared, on foot, out of the rain, wearing a hooded sweatshirt, a pair of khakis, and white tennis sneakers, his demeanor placid yet purposeful.

"Did my call wake you?"

"No," I said. "How about you? Were you in bed?"

"More or less," he said, going on to explain that he'd gotten a heads-up call earlier in the evening while having dinner with his wife and friends at his home in Westchester, and therefore had one foot out the door all night.

"But I should be back home well before noon," he said cheerfully as we started through the front lobby toward the bank of main elevators. "We're only going to Newark."

We rode up to the sixth floor, then passed through a network of corridors to the surgeon's locker room. A second-year intern named Joe Slater, Dr. Rosen's assistant for the harvest, was already changing. Dr. Rosen grabbed me a set of scrubs from a tall metal cabinet. Following his lead, I put on just the sickly blue pullover shirt and matching drawstring pants, saving the separate packeted head and shoe coverings for our pre-surgical scrub at the operating room in Newark.

In a narrow hallway outside the locker room, Slater grabbed a blue Gott picnic cooler, placed it on a wheeled cart and held open the lid as Dr. Rosen tossed in large, clear bags of saline solution and of cardioplegia fluid, the chemical compound with high concentrations of potassium used to stop the donor's heart just before it is cut out.

Rosen and I followed behind as Slater wheeled the cooler down the hallway through a side door that opened onto one of the nurses' stations on the sixth-floor's cardiac intensive care wing. Tucked within a little alcove was an automatic ice machine. Dr. Slater began to fill the cooler, the crunching scoops resounding through the hushed, low-lit ward. A couple of nurses sat sipping coffee before a bank of weakly cresting heartbeats, while some of the ward's end-stage heart-failure patients shuffled soundlessly past in the background, doing slow, restless laps with their wheeled intravenous poles, unable to sleep for fear of their unpumped body fluids gathering in their lungs.

"Pole people," they call themselves, or "pole-pushers," heart-transplant candidates whose condition has become so dire they need a constant infusion of drugs to keep their hearts beating until a suitable donor heart can be found. I'd been up to their ward once before, following Dr. Michler some weeks earlier on his afternoon rounds.

Room after room, it was the same, people literally at the end of their tether, where all those distinctions by which we habitually assess ourselves and others, things like personality, appearance, sense of humor, have been discarded, life dwindled down now to its lowest common denominator: the next few breaths and the chance, however remote, of a last bit of lifesaving news.

It is a tenuous, topsy-turvy existence, life with a nearly dead heart, another of those little twilit clearings made by the hands of modern medicine between living and what would have meant, until very recently, certain death. You often feel breathless exhaustion, even at rest. You gain weight without eating, your extremities bloating for lack of a strong enough circulation. You have to sleep sitting up. Your hands and feet are always cold. Your complexion is gray. Your speech is perforated with urgent, shallow breaths. Daylight looks paler to you, and yet in the general diminishment of all your body's functions, oddly specific ones rally to compensate—your sense of smell, for one. Otherwise faint, apparitional days are suddenly cluttered by too-strong odors; perfumes and colognes seem more substantive to you than their wearers.

All the old, dreaded hospital associations came flooding back as I followed Dr. Michler through the cardiac ward, thinking the whole time about my father, and the fact that, had his heart somehow been able to hang on for another five or so years, he would have likely become one of the pole people, a member of their eerily select society.

Dr. Michler seemed to drift in his neon-green clogs above the ward's shiny linoleum floors that day, held aloft at once by his patients' beseeching gazes and his own awareness of them. Doctors have an unreal, outsized aura as it is, perhaps because we see them as we do, only when we most need to be delivered from what we want to think about least. But the mere appearance of Dr. Michler at a patient's door was enough to stir whatever humors remained in their ebbing blood's tide: smiles coming now to sallow, formerly fallen faces; once-supine shapes sitting suddenly forward; and, here and there, even a discernible edge in some of the responses to Dr. Michler's airily soft, occasionally unctuous expressions of empathy.

"Yes, I know . . ." he said, lightly pinching with a thumb and forefinger the covered big toe of a young woman named Carol Godwin, a thirty-nine-year-old single mother whose heart had contracted a virus two years earlier. No one could say which one exactly, tuberculosis, perhaps, or a common cold, but it did irreparable damage to the heart muscle. On the transplant waiting list for well over a year, Carol had been living at home on a variety of medications, but in the excitement of her recent wedding engagement, and her preparations for the upcoming Christmas holidays, her heart began to fail completely and she was forced to come to the hospital.

"I know . . ." Dr. Michler kept saying, bearing a stiffly beatific smile. "You sit tight. We'll get a call before long."

"Where am I going to go?" Carol asked, shaking back a full head of blond hair and casting a disgusted glance at the IV rigging, or telemetry pole, just beside her bed. A six-foot-high metal pole on a tripod wheel base, it supports a computerized telemetry box that monitors the steady flow of a drug called dobutamine directly into her heart, forcing it to beat.

"I hate this thing," she muttered, a pretty woman, as far as I could tell, having to extrapolate her former appearance from the sorry one that her failing heart had fashioned: pale, bloated cheeks, sunken eyes, a weak, plaintive voice.

"It's kind of sad all around when you think about it," Carol went on, the room going quiet as she stared out her window through a light snow toward the Hudson River and the George Washington Bridge. "I mean this is a sick way to get better, someone having to die for me to get a heart. But then I'm not living either, am I? This isn't living. I'm kind of backed into a corner here."

It was near dusk by the time we reached the door of the last patient on the ward that day. "I think I know where to find him," Dr. Michler said, turning at the sight of Bill Tierney's empty bed, escorting me now down another long hallway.

"He spends a lot of his time in the solarium."

Thoughts and blood: winter of 1978, my leg cast in the same frozen, drifty white as the surrounding Chicago suburbs. I'm wheeling my way one evening down a pale green hospital corridor, where night brings only the next shift of nurses to your room, changing the sheets before you can settle into them, leaving behind little white plastic replicas of themselves on the walls, light spreading out from under their skirts.

I take the elevator up to the cardiac wing, wheel down

another hallway, arrive at the door to my father's room. His bed is empty, and now I'm wheeling frantically toward the nurses' station, thinking that the "slight arrhythmia" for which he'd been admitted just two days after my knee surgery had now gone completely awry, and they couldn't get it righted again, and I've come too late.

"He's just down the way," the nurse says. "He's in the solarium."

I go to the end of the hall, where there is a large bay window. A young man is sitting beside a pot of dried plants, his hands wrapped in gauze, the right hand barely wielding a pen as he tries to write out a card. Across from him, my father sits in a wheelchair, watching the evening news.

"He'd rather you hadn't," I'm thinking to myself even as I'm wheeling toward him, a mere pawn now of that moment's peculiarity, mindlessly drawn, yet again, toward completing it. "He'd rather that you left him alone, stayed away until he's back on his feet and able to come see you instead."

I pull alongside him. He looks over at me, winks. We sit there together, watching TV in silence. Somehow that deep, centrifugal shyness that often overcomes fathers and sons seems completely natural now: a kind of shared mortification over our mutual excerption from life's story.

"A bunch of us pole-pushers like to gather here," Bill Tierney told me as he sat before the cardiac wing's sixth-floor solarium window, looking out over the lighted frets of upper Manhattan's streets just beginning to take hold in the darkness. "Especially on the rainy nights. We'll all sit here, watch out the window, and . . ." a queer smile crossed his face as he turned to make sure that Dr. Michler, who'd just started back to his office to make some phone calls, was out of earshot, ". . . you know, wait for stuff to happen."

Bill, I soon learned, was waiting for his second donor heart. The first one, implanted seven years earlier when he was only twenty-six years old, was now succumbing to the coronary artery disease, which, for reasons still unclear to doctors, eventually afflicts nearly all heart transplant recipients.

"And why did you need a transplant to begin with?" I asked.

"Well," he began, a tall, broad-shouldered figure with a face that might be construed as hard-edged, mean, if not for the softening creases at the corners of his eyes, "you've probably never heard of it. It's something doctors call dynamite heart."

I was too embarrassed to tell him that I had, in fact, heard of it, a rare condition resulting from overexposure to nitroglycerin, which, as Bill would go on to explain, he incurred while working on a highway construction crew, the daily blasting through rock causing his blood vessels suddenly to dilate and then, in his off hours, contract again, severely restricting his heart's blood supply.

The paths to heartbreak, I have learned, are nearly as varied as those nature would take to make the heart. As Bill was telling his story, I found myself reviewing in my mind the many heart maladies I've read about over the years, all the mind-twisting terms I've looked up, cross-referencing and re-referencing them countless times, as though by knowing the diseases in advance and being able to conjure their various symptoms, I might somehow construct some secret inner shield against them.

People hear and talk all the time about the "heart attack," think of it as the one all-encompassing form of cardiac illness. This is particularly odd considering that a heart attack is most often not even the result of an inherent problem with the heart itself, but with the arteries that supply the heart its own blood: a relatively straightforward plumbing problem, which, diagnosed in time, can be dealt with through diet, medication, and, as a last resort, bypass surgery. There isn't even a listing in any of my medical dictionaries for "heart attack."

"Heart failure," on the other hand, occupies whole columns, maladies as variable and insidious as the metaphoric organ has long been described as being with its own warring impulses, its propensity for deceit and betrayal. Indeed, the very wording of my 1948 Stedman's definition for "heart failure"—however archaic it may be medically—is evocative of the heart's inwardly riled, deeply ambivalent nature: "Inability of the heart muscle to maintain the circulation, its embarrassment and exhaustion being due to some disturbance in the normal balance between the propulsive force and the resistance to be overcome."

In each heart's failure, yours, mine, at one time, two disparate motions, two opposing impulses, a self-contained, autoerotic tug of wills, such "underlying metabolic defects" as a more current medical dictionary calls them, disrupting all those precise inner body gradients and pressure fields that Neal Epstein had told me about: the literal heart of our every living moment succumbing now, in varying ways, to life's own longing for entropy.

There is the dilated and the hypertrophic form of cardiomyopathy—a general term for disease of the heart muscle—and something known as restrictive cardiomyopathy, in which the heart's ventricular walls become excessively rigid, preventing the chambers from adequately filling with blood. These conditions are often "idiopathic." Sometimes they are the result of a genetic flaw, and other times the secondary result of another disorder, like high blood pressure, or valvular disease, or what's known as cardiac ischemia—insufficient blood flow to the heart because of the very blockages in the blood vessels that cause heart attacks. Heart failure, in fact, is often the end result of a number of heart attacks.

A variety of inflammatory diseases affecting the pericardium—the moist, fibrous, two-layer sac that surrounds and protects the heart—can also result in heart failure. You can, for example, develop "bread-and-butter heart," in which the inflamed pericardium is covered by a thick, fibrous exudate resembling, as Stedman's describes it, what is "produced by separating two slices of buttered bread." Or you might have a "hairy heart," in which that same fibrous exudate takes on a fuller, "shaggy" appearance. Or a "frosted heart," in which the pericardium develops a thick white coat like cake icing. Or "armored heart," in which the pericardium calcifies as a result of chronic inflammation.

An infant can be born with a hole in its heart, or with a "three-chambered heart." Pregnant women sometimes develop, for unknown reasons, irreversible heart failure or "peripartum cardiomyopathy" in the last month of pregnancy or the first few months after delivery.

A heart can be done in by environmental and behavioral factors. "Beer heart" and "alcohol heart" are common dilated cardiomyopathies, an extreme distention and thinning of the heart muscle, resulting from the toxic effects of excessive drinking, my great uncle Henry's probable pathology. "Beriberi heart" is a form of cardiomyopathy caused by vitamin-B deficiency. A "tiger" or "tabby-cat" or "tiger-lily" heart is one striped and crippled by fatty deposits, whereas an "irritable" or "soldier's" heart is a neurocirculatory weakness marked by rapid pulse, shortness of breath, and fatigue.

"Irritable heart" is most often observed in soldiers in active war service, but it also obtains among the civilian population, and is a malady that, like my long-ago bouts of "heart hurry," belongs to that same beguilingly labeled subset of heart disorders known as "cardiac neurosis."

Cardiac, *adj.* (from the Greek *cardia*, and the Latin *cor*, both derived from the Sanskrit root *kurd*, signifying at once an even, rhythmic swinging or jumping, and a twitching, trembling, hesitating motion): "Of, near, or pertaining to the heart . . ."

Neurosis, *n.*: "Any of various functional disorders of the mind or emotions, without obvious organic lesion or change . . ."

Still, the term "cardiac neurosis" refers not, as one might first assume, to a neurotic obsession with, or phobia about, the heart, but to the profound organic changes suffered by the heart as a result of the brain's upheavals: the heart, in other words, being driven out of its own mind by the mind.

Thoughts and blood, time and memory, coming full circle, coalescing only in an evanescent heartbeat: early spring of 1978. I am completely on my own now: the leg cast removed, the Holter Monitor returned, my heart's divorce finalized.

I am on my way home to my newly rented apartment on the Lower East Side of Manhattan, walking along St. Mark's Place, trying not to think about my heart. My hands still reek of blood, that high-pitched, vaguely metallic taint that no amount of lemon and soap could cut after another full day of butchering, slicing whole sides of beef down into dinner-sized steaks, Beethoven's piano sonatas sounding from the portable tape deck my father had gotten for me wholesale from one of his buyers before I left Chicago that winter, its dials and buttons all caked by then in animal fat.

Near the corner of St. Mark's Place and Second Avenue, I stop into a used-book shop. There, in a back bin, is a collection of verse by the Spanish poet Vicente Aleixandre. I start flipping through its pages, my eyes instantly drawn to the first line I see containing the *H* word.

"I want to know," it begins, "if the heart is a rainstorm or a riverbank."

This, it occurs to me, gets closest to the true nature of our hearts and of our deep awareness of them. It's a plea the meaning and urgency of which I couldn't have known prior to that winter when I first felt my own heart's enactment of the emotional extremes for which "rainstorm" and "riverbank" are metaphors: the torrid pace of panic and the resonant beating, the "even, rhythmic swinging," of my resolve to overcome it; the fits and starts, the "twitching" and "hesitating" of our passions, and the even, measured pace of our periods of reflection upon them.

My winter of heartbreak had, in essence, delivered me to the shockingly banal conclusion that the heart is the seat of our emotions, or is, at least, the place where our emotions seem most heavily to sit. That the heart is not only the mirror but the subtle mime of our ever-shifting physical and psychological states.

That behind Alcixandre's plea, or our most clichéd Hallmark card phrases (my heartfelt joy or sympathy; my heart leaps, longs, aches or breaks for you), there are physical correlatives, feelings in the body caused by or related to the varying actions of the heart. Over the centuries of thinking about and expressing our emotions, we have not arbitrarily or fancifully assigned them to the heart. We have, in effect, been eavesdropping upon and trying to catch up with and rephrase in our own words the heart's far more demonstrative responses to the upheavals of our brains.

"You did actually use one word I like," I remember a very wary Neal Epstein saying when he first heard my thoughts on this matter, half believing himself, I think, that I was trying to set science back a couple of millennia by reigniting that ancient debate between the cardiocentric

Aristotelians and craniocentric Platonists over the precise location of our true essence.

"The word 'mime.' It's the communication between the heart and the brain that we still don't fully understand. That's what needs to be spoken about."

It is, in essence, our first and oldest conversation, and one to which the ancients, for all they didn't know or understand about our body's inner workings, were most acutely attuned, back when myth and science were—just as our hearts and brains are at the very earliest stages of our development in the womb—closely aligned with one another, essentially one: a way of explaining phenomena and exacting intuitive truths from it—truths that our most recent scientific discoveries about the heart are now prompting us to rediscover.

Among the many studies I'd gathered from newspapers and medical journals in the years since my "divorce," one, devoted to the patterns of the heartbeat, found that contrary to the established notion of the strong heart being a consistently metronomic, lub-dubbing pump, the healthiest hearts, in fact, beat rather chaotically, in constant and subtle concert with our daily physical and emotional stimuli: findings that would hardly have surprised ancient Chinese healers, who, as long ago as 2600 B.C., had documented fifty-one different varieties of pulse.

Or early Greek and Roman physicians. They, of course, had a number of things wrong in the absence of the knowledge of internal biology yielded by modern instrumentation and accrued research. The Greeks, for example, believed air or *pneuma* flowed in the arteries (thus the name, from the Greek *aer* [air] + *terein* [to keep]) and thought blood to be a liver-borne distillate of ingested food that got heated and burned up within us like fuel in a furnace. Plato, meanwhile, posited in his *Timaeus* that the body is something the gods "put together to prevent the head from rolling about on the earth, unable to get over or out of its many heights and hollows . . ."

Still, Greek physician/philosophers intuitively surmised the central and pivotal existence of atoms, and, when it came to our bodies, posited as well the metaphorically sound notion of the heart as the governing force of the so-called humours, the respective physical correlatives of those humors—blood, phlegm, bile, and black bile—being the fluids that comprise much of the inward sea in which we are kept adrift by our hearts at varying depths, depending on the heart's varying rhythms and output, a variety that Galen—the famed second-century-A.D. Greek scholar, researcher, surgeon, philosopher, logician, herbalist, and personal physician to the emperor Marcus Aurelius—had figured to be closer to twenty-seven varieties of pulse, among them the "antlike," "wormlike," and "wavelike."

"In anger," wrote Galen, "the pulse is deep, large, vigorous, quick, and frequent. In pleasure it is large and sparse, but no different in vigor. In grief it is small, slow, faint, and sparse. In sudden, violent cases of fear it is quick, tremulous, irregular, and uneven; in cases where fear has been present for a long time, it is the same as in cases of grief."

At the end of my stay in London, I happened upon a

series of studies at the Wellcome Institute Library about the effects of grief and other forms of emotional stress on the heart. In one such study it was determined that the cardiac arrest suffered by nine Southeast Asian refugees—all women with no previous indication of heart disease—was caused by the emotional trauma they had experienced during the war in Vietnam and their subsequent displacement from home and family.

Another report, conducted at the University of California's Irvine Medical Center, determined that severe emotional stress can cause sudden and sometimes fatal arrhythmias—disruptions of the normal flow of electrical impulses through the body's sympathetic and parasympathetic nervous systems, which regulate the beating of the heart.

"People," the study's head physician concluded in the *Journal of the American Medical Association*, "do suffer heartbreak."

Sympathy, in other words, has its own physical infrastructure, at the very center of which stands the heart, and right then a clear picture began to coalesce in my mind of what actually did happen between my father and me that long-ago night in the winter of 1978, the night when, as I now saw it, my heart, against my own conscious will, literally went out to him.

"I sometimes envy writers," Bill Tierney said to me, pulling himself up from the solarium sofa with the help of his IV pole. "Writing is so portable, a pen, paper, and your thoughts. I started taking up sculpture after my first transplant. You need lots of material and lots of space. But I like it. It soothes me somehow."

"What kind of stuff do you make?" I asked, as he started shuffling back out into the hallway.

"Mostly," he said, "big metal trees."

The last time I'd see Bill and the other pole people was at one of their regular Monday-afternoon education classes and support-group meetings. Heart-transplant patients have a wholly independent infrastructure to help them deal with the ramifications of their upcoming surgery: specially trained nurses, social workers, a team of psychiatrists; weekly meetings and therapy sessions. They even have their own weekly newsletter titled Heart Talk.

A number of vital organs are transplanted these days. Their respective harvest teams were all racing that night to the side of the same donor in Newark whose heart our team would be harvesting. And yet the intended recipients of their bounty don't read copies of Liver Lessons or Kidney News. As much as we've learned about the heart, as clear as we all seem to be now about its proper place in the organ hierarchy, it still has this deep hold on our psyches, still has us uttering phrases like "I'm speaking to you now from the heart," or "I know this in my heart." Not merely out of habit or a sense of nostalgia, but from an ongoing sense of awe. We feel a kind of deep, subconscious fealty to the heart not merely as the living core of us, but as the palpable reminder of our own primal, pre-thinking selves. It was, in a sense, their lives' first and oldest love that the patients at that Monday-afternoon meeting were being

prepared to part with, their mind's long-trusted counterpart, the most basic measure any of us have of our being:
the part that both gives us life and seems to take it the
hardest.

The meeting was held in a conference room directly
opposite the nurse's station where we were filling the picnic cooler with ice that night. There were eight patients in
attendance, seated about a long oval table, IV poles at their
sides. Carol Godwin and Bill Tierney were seated next to
one another.

"Did you hear about Watson?" I heard Bill asking
Carol as he was settling into his seat. He was referring to
Tom Watson, one of their fellow pole-pushers. "He got his
heart last night."

"I know," Carol said. "Just after midnight. Hear he's
doing great."

A social worker named Ann Lawler opened the meeting
by introducing the transplant coordinator, Kim Hammond.
Hammond was there to discuss the risk of postoperative
infections that transplant recipients face because of the
high doses of immunosuppressant drugs they must take
to ward off their body's rejection of their new hearts. She
began listing the types of things to be avoided when they
leave the hospital: flowers, construction debris, humidi-
fiers, potato-skin spores, cat litter boxes.

"Even a child's tears can be dangerous," Hammond said,
"if that child has just been immunized with a live virus."

Everyone was taking notes. It was as though they were
a group of tourists preparing for their first trip to a for-

eign country, trying to establish what if anything will be familiar to them when they get there. Carol asked about sushi and raw clams, her favorite foods.

"Absolutely not!" Hammond answered. "No raw foods."

Someone else asked about sunny-side-up eggs. A long discussion ensued about sunny-side-up eggs. I kept imagining my father there, kicking up a huge fuss over his beloved undercooked burgers and steaks, and his absolute favorite, steak tartare, the way he himself insisted on making it: raw chopped meat heaped upon on a slice of rye, with a raw egg and a dense coat of salt and pepper on top. I could just see him, sitting there among his fellow pole-pushers, seriously weighing the benefits of a prolonged but drastically denuded life with someone else's heart versus a horribly foreshortened and yet fully indulged existence with his dying, natural one.

"No," Hammond insisted. "Only fully cooked eggs."

Just then a small, middle-aged woman with close-cropped black hair came into the room. She took a seat at the table and introduced herself as Mrs. Tom Watson. Everyone applauded.

"His color is great," she said, excitedly, almost breathlessly. "It's great all over. Ten weeks he waited in this hospital. He was very sick. Now he's sharp. He's been talking all day. He won't shut up. Never in your wildest dreams can you imagine."

There was more applause. Then she held out a piece of paper.

"Tom put this in my hand. He said this is the number to

play in the lottery—8117. Eight months, eleven days, seven hours. That's how long he waited to get his new heart."

The comment fell strangely flat. Everyone's fate was by then so entirely dependent upon the advances of modern medicine that there seemed now to be no allowance, no room in their dying hearts, for things like luck and super-stition. Those would have to wait for later, for when they got back to their life's story.

Ann Lawler thanked Mrs. Watson for coming, and then turned the meeting over to the patients, asking them to discuss their feelings. No one said a word.

"It's very upsetting," Carol finally announced. She paused a moment, shook back her hair.

"I want everything yesterday. I fight with the nurses. I'm a real pain in the ass."

Everyone was laughing now.

"It's this waiting. And the fear of the surgery. You can't keep your focus. You try to watch a movie, and then your mind starts to wander. Why did this happen to me? What will it be like afterwards?"

There were solemn nods around the room. For a mo-ment, with everyone sitting there beside their IV poles, the transparent tubes dangling down, they looked like a group of sickly paratroopers readying themselves for a jump into the unknown, into some uncharted place on the far side of their native hearts. Some of the IV poles had little per-sonal mementos attached to the tops. I hadn't noticed them until then: Carol's wooden heart trimmed with painted flowers and a pink ribbon around the words "Friends are angels in disguise"; Bill Wilson's dangling mobile, one I figured he must have sculpted himself: a gray metal hand, with a bright red heart slowly turning within the hollowed-out palm.

I had asked Bill in the solarium that evening, he being the only one who could speak to the matter, a version of the very question Carol had just posed, about what it is like afterwards, how it feels to have another person's heart, to be—and he laughed when I inadvertently phrased it this way—a man after his own heart.

The question was not rooted in old heart mythology. A transplanted heart, as I was to learn in my conversations with Dr. Michler, can be attached to all of a recipient's vital veins and arteries, but not to his nerves, neither the parasympathetic vagal nerve, which, like an engine at idle, keeps our hearts between sixty and eighty beats a minute when we're at rest, nor the sympathetic nerves, which make for the subtle varieties of pulse that the ancients wrote of, alternately speeding or slowing the heart during periods of exercise and cool-down, or miming the brain's everclashing spells of emotion. In the process of receiving a new heart, in other words, a transplantee is also effectively severed from that lifelong to-and-fro between the heart and the brain that first gave rise to all of the old emotional clichés: "My heart leaps up, " "Be still my heart," and on and on.

"You do feel a difference," Bill said, pausing to recall what his donor heart felt like when it was still healthy. He told me that it went at a constant rate of about 100 beats a minute, and that when he exercised he had to both start and ease off gradually, allowing his brain to recognize what he was doing and then either signal the release of blood-borne hormones like adrenaline to get his heart rate up, or, after he'd stopped exercising, to cut that signal off and let his heart slow back down.

"But another thing I noticed," he said after a moment, "is that it has become a lot harder to scare me. Like if a friend sneaks up from behind and says 'Boo!' I just kind of quietly turn and say, 'What are you doing?'"

Ann Lawler had begun to draw the meeting to a close that day, when the door to the conference room opened. A hospital custodian entered with a small stepladder. He got up on the very top step, strained to remove a ceiling light panel, then came down with a spent fluorescent bulb and left. As the meeting was letting out, I noticed Carol and Bill and a few of the others standing together by the door, laughing. Once the room had fully cleared, I asked Ann Lawler what the joke was.

"When the custodian was up on that ladder," she said, "they were all thinking the same thing. If that guy falls, hits his head, and dies, who gets his heart?"

Chapter 14

A PLAIN WHITE SEDAN, WITH THE WORDS ORGAN TRANSPLANT UNIT written on the side, was waiting out in front of the Milstein Hospital building when we got back downstairs, the three of us wheeling our way with the cooler through the nearly deserted front lobby. I thought it odd that there wasn't some other, special side door or back door for this sort of undertaking, that we were rushing at 2:30 A.M., with our ice-packed Gott cooler, past a darkened gift shop and a few story-exiled wanderers with special visitor passes strung around their necks. It felt as though we were a trio of overeager fisherman, determined to get out on the water early.

Our driver, a moonlighting New York City fireman named John Bezares, was waiting for us by the front door. He opened the car's trunk, put the cooler inside, and we were off, Dr. Rosen riding up front, Dr. Slater and I in the back. No sirens. No screeching tires. No running through red lights. It's not that we weren't in a hurry. We were, as I was about to learn, in a peculiarly calculated kind of hurry— part of a little delay game that Dr. Rosen said a heart-harvest team sometimes needs to play with the other harvesters.

The heart is first in the hierarchy of organ extraction because of its relatively limited ischemic time and the fact that there is a patient concurrently being prepped on the operating table at the other end.

"We're the only ones who can lose both the organ and the patient," Dr. Slater explained.

The liver comes second. A liver can survive about

twenty-four hours on ice before it's implanted. Next go the kidneys and then the pancreas. After those follow the eyes, skin, and bone marrow, those all being items that, as Dr. Rosen put it, "get banked."

Still, the precise moment of excising the heart has to be carefully coordinated with the readiness of the patient to receive it, requiring constant communication between the harvesters and the transplant surgeons in order to ensure that our "catch" spends the least amount of time as possible out of its element.

"We'll probably have to stall a bit longer than usual when we get to the hospital," Dr. Rosen said as we headed for the George Washington Bridge. "The match for the heart we're getting is a Status 2. He hasn't even gotten to the hospital yet."

I gave a quick glance back toward the hospital building, knowing, now, that this wasn't to be Bill's or Carol's night, or that of any of the others probably standing up there at this moment in the solarium window, beside their IV poles. An unmistakable row of silhouettes, like a vaude-ville stage set of slumping drunks against lampposts, looking out over the thrum and siren-thread of a wet city night, trying to parse the random, rain-blurred lines of traffic for some personal signal, wondering if ours or the next set of car lights might not be remotely related to their cause: the last viable match being borne through the damp by an unseen hand toward their hope's last, carefully guarded stack of kindling.

We all fell silent as our car pulled up onto the upper level of the George Washington Bridge, its bright stanchion lights setting off the Hudson River's gray, rain-roiled waters far below.

"He drowned . . ." I remember a woman's voice on the telephone saying, "The doctor said he drowned in . . ."

I'm standing in my upstairs neighbor's apartment in Houston, Texas. It's far too early in the morning, February 1, 1980. There is a metal sculpture on the table beside her telephone, a tiny human figure leading an equally diminutive deer through a tall stand of bare, silvery trees. Outside her apartment window, I can see that the rain has stopped and daylight is nearly in.

I remember this because I had just been watching for it out my own window. Had been downstairs in my ground-floor apartment, just moments before the knock came on my door, sitting up before dawn at my kitchen table, watching out the window the rain finally stopping and the day coming in. I had gotten very little sleep that night, the Southern Pacific freight trains clamoring all hours through the switch yards at the far end of Polk Avenue, their whistles sounding especially plaintive, mournful.

More portent. Always pieced together in retrospect: those train whistles and that lone barking dog, the one that was tied each night to the pole of an open-air carport just up the block from my apartment building. I'd gone down there one night earlier that same week and knocked on the house door nearest the carport. Getting no answer,

I decided to post a note on the pole to which the dog was tied, asking its owner to keep it inside. The next day the dog was still there along with my note, flipped over now, bearing a new message, scribbled in pencil:

"This dog no bother you," it read, "as you must talk, a dog must bark."

I'm sitting there at my kitchen table, before dawn, staring out the window, trying to prepare for the writing class I'm scheduled to teach that morning at the University of Houston. Behind the row of houses across from me, the freight trains are still mining the dark for daylight. Factory workers in blue jeans, T-shirts, and cowboy boots are milling about on their front lawns, waiting for rides. The rain has slowed to a drifting mist, and the floodwaters that nearly swept me away the previous day have almost fully receded again beneath the sewer grates.

Houston is prone to flash floods. The city was built on a vast swamp, and is sinking now over an inch a year. Water regularly bubbles up out of sewers on the driest of days. In my neighborhood in particular—a rundown, blue-collar barrio on the industrial east side of Houston, the only place that I could afford to live within walking distance of the U. of H. campus—the city's drainage system, such as it is, was in dire need of repair.

Still, the storm had just started the previous afternoon when I walked over to the supermarket across from my apartment building on the far side of Polk Avenue. By the time I'd made my way through the checkout line, other shoppers were all clustered at the front of the store, having chosen to wait out the first deluge. I, however, decided to make a break.

A bag of groceries in each arm, I raced across the parking lot and down the entrance ramp onto Polk Avenue, where my legs now felt thick and slow. I looked down to see the bottom of both grocery bags coming undone in the newly formed Polk River, eggs, milk, bread, and the rest, bobbing off under a nearby train trestle. Then I left my feet and began flailing, furiously, for the far shores of Polk Avenue.

You just swim. They've got this all figured, you're thinking, the adults. They know the way out. The balloon man never lets go. You dive down and make the mad, slime-wall grab toward the light and the coarseness. You have no idea how long it is taking. Once under water, there is even a bit of ease within the urgency, a faint call, playing along that one fish-gill vestige we all retain within our inner ear, a call to stay, and then someone's waiting hands are there, pulling you up into air and the news of him.

The rain had stopped for a moment when I got back to my apartment, and the sun briefly appeared. I went to my kitchen window. Water was slapping up against the base of the sill, the waves sent in my direction by a group of neighborhood children, who were trying to ride their bikes through the chest-high swells, some of them waving at me above the sound of their water-muffled handlebar bells.

Somewhere within the piles of papers strewn across the wooden cable-spool table in my Houston apartment's living room early that following morning, February 1—my clothes still drying on hangers fixed to the neck of the light fixture above the apartment's single gas wall heater—was my father's latest and, it would turn out, last correspondence. He'd sent it only a week before, the same single, folded page of unlined, premium-bond stationery, crammed with the perfectly aligned penmanship:

> Dear Chuck,
>
> Enclosed please find a copy of the form pertaining to your last New York City apartment's broken lease. I don't know how it got to this address, but I thought you should have it for your records. All is well here. Your mother and I are planning a trip by train to St. Louis on the morning of February 1st to see old friends. We'll only be gone for a few days. Weather here is still very cold.
>
> > Love,
> > Dad
>
> P.S. Please let us know your new number if you do manage to have a phone installed.

Resting on that same table, atop the satchel of books I was intending to cart off to class that morning, was the letter I had managed to write before going to bed that night, one of the few I'd ever written him that was not a protracted segue to another request for money. It was, for the most part, about his favorite subject, the weather, that day's violent storm, the flash flood, the desperate swim home, although at its close, I do remember conceding to him, half-seriously, perhaps, the foolishness of my ever having left butchery to try to "take flight" in a city that was sinking over an inch a year.

I am seated at the kitchen table. The water has fully receded by now. The rain and the darkness are almost gone, just some gray beads left on the air, the neighborhood's stray dogs trying to swallow them, like fish. A pair of truck lights pull up to the house across from me, the horn sounding three times. A man comes half-dressed to the doorway, stairs of shadow and light behind him, puts on a straw hat, whistles, and starts across the wet lawn. Down the way, the tied dog suddenly goes quiet. In the low, waxen magnolias and live oaks, I can almost hear the birds unwinding their heads from the rain. Somewhere in Houston's thick, subtropical air, there is even the suggestion of a breeze.

A knock comes at the door. My upstairs neighbor. I'd given my parents her number in case of emergencies, a young woman from Natchez, Mississippi, a law-school graduate, looking for work. She's at my door, saying "Hey," very softly, her deep drawl rendering even that one syllable as three. "Someone's on the phone, cry-uhn . . ."

And now I'm standing upstairs in her apartment, looking down at a tiny figure leading a deer through a stand of metal trees. There's a voice—my mother's, perhaps, or it may be one of my sisters', mindlessly repeating, "He drowned. He drowned in bed . . ." words that somebody in my family must have overheard a frustrated ambulance attendant or an emergency room physician say.

Somewhere in the course of that first conversation I

had with Neal Epstein in his lab office years ago, he did eventually come up with an example of a multicellular life-form on this planet that doesn't possess a heart.

"I suspect there's something in the ocean," I remember him saying, slowly spinning toward me in his seat. "A sponge!" he cried, clearly annoyed with himself. "A sponge, for God's sake. A sponge is all the same cell, but it's the same cell multiplied. Now, what does it use for a pump? The ocean. The ocean is the heart of a sponge. The sponge has been designed to live within a huge pump, bathing it in nutrients, washing away its waste. Now, you'll notice that the salinity of the ocean is not that different from what we're carrying around in our bodies. So somehow, when life came out of the ocean, it had to figure out a way to carry around its own bathing ocean and circulate it."

Ever since, I've thought of each of us, you and me, as our own pinched-off, ambulatory seas, ones in which only our hearts keep us buoyed. When our hearts fail, we can no longer tread our own bodies of water, and we drown.

Chapter 15

OUR DRIVER, JOHN BEZARES, WOUND UP TAKING the wrong Newark exit that night: three scrub-clad men and a New York City fireman in an Organ Transplant Unit vehicle, racing up and down the city's wet, deserted streets, asking whomever we could find for directions.

I had originally thought it a really bad joke. Then I saw the mortification in Bezares's eyes, and the increasingly nervous fidgeting of both Dr. Rosen and Dr. Slater as the first three people we stopped were unable to direct us to the right inner-city hospital, one that, as part of the terms of my "arrangement" with Dr. Michler and Dr. Rosen, I had to agree to keep anonymous for the sake of the donor's family.

Heart-transplant recipients often go out of their way to learn the identity of their donors, not only to thank his or her relatives, but to know more about "who" it is that they now have living inside of them, whose vital half of a whole other lifelong conversation they have admitted into their own.

Many recipients of transplanted hearts claim to have assumed certain aspects of their donors' personalities. This is not, it seems, merely willful heart mythologiz-ing on their part. Our hearts are now thought to have, in ways we are just beginning to measure, their own separate energy fields within the larger field of bodily energy that the heart itself orchestrates: that dynamic suspension of concentrically bound fluids—from the smallest cells to the largest, outer cask of the skin—through which each heart-beat thrums and warbles like a just-struck tuning fork.

The heart, it seems, retains the imprint of those rever-
berations in the form of what is known as cellular memory. Just as each heart contains the echo of the very first
heart's formation, it retains as well, within each of its
cells—from the Latin *cellula*, meaning "small chamber"—
the imprinted cellular memory of the very life it informs.
In fact, recent research in the field of neurocardiology
has revealed that the heart has its own built-in nervous
system: cells that synthesize and release neurochemicals
like dopamine and certain hormones once thought to be
the exclusive province of the brain. Poe's "Tell-Tale Heart,"
I thought to myself as the hospital finally appeared in our
headlights, calling us to itself.

Our impromptu, wee-hours tour of downtown Newark
that night wound up fitting quite nicely into Dr. Rosen's
little delay game. Upon pulling into the emergency-room
parking lot, Dr. Rosen went to a wall phone just inside the
ER entrance doors and called Columbia-Presbyterian. The
transplant recipient had just arrived at the hospital and
was only now being prepped for surgery.

Dr. Slater, meanwhile, had gone up a back hallway and
returned moments later with a wheel cart. He brought it
out to the car. I put the cooler on it. As we started off in
the direction of the operating room, I turned back briefly
to see a very relieved-looking John Bezares settling back
into his driver's seat with his coffee thermos and a copy of
the *Daily News*.

Dr. Slater parked the cart in a hallway just outside the

operating room. We put on our shoe coverings and surgical caps, went into an adjacent scrubbing room, then reemerged and headed for the OR.

I must have hesitated upon first entering, nearly blowing my cover. Dr. Rosen had gone in first. I remember being directly behind him, and then, somehow, I was not, standing there alone in the brash lights, watching Drs. Rosen and Slater in a far corner, opening the blue Gott picnic cooler.

I couldn't see the operating table, just the other harvest-team members—liver, kidney, pancreas—pressed elbow to elbow, working. Some of the OR attendants were staring at me. One started to approach. I felt the warm press of a hand on my right arm. It was Dr. Slater, leading me further in, taking me to my "proper" place at the very head of the operating table.

"He is here," Dr. Slater announced, "to observe."

I was standing directly over the table now, my hands dangling just off the top edge on either side of the donor's covered head. She was lying there before me, split open from the shoulders to the waist: a glistening, multi-shaded inscape of organs. But for the lungs—spongy pink, diaphanous—rising and falling with the whooshing clicks of an adjacent respirator, all that moved was her beating heart.

I remember my body making one brief, dizzying pitch forward, and then, like the sudden flip of a focus knob, a compensatory lurch back. And then the mind does this, the layman's mind, at least. It withdraws, pulls, like a just-brushed anemone, all of its feelers in. Goes toward the lowest common denominator of thought and feeling, starts to repeat itself mindlessly: *She is lying here before me, split open from the shoulders to the waist. But for the lungs, all that moves is* . . .

My subject. My source. My longtime nemesis: deep crimson, looking far more like a throbbing mango than the all-too-familiar heart symbol, that bulbous-topped, up-raised, come-hither-ass of a shape as old and ubiquitous as symbol-making itself, one of its earliest known appearances found on the wall of a cave in Asturias, Spain, a prehistoric drawing, circa 10,000 B.C., of a woolly mammoth, its dark, austere outline further setting off the cordate red dot set within.

It's a shape that seems to have no definitive origin. Some have attributed it to the ivy plant, the heart-shaped evergreen leaves of which made ivy an early mythological symbol of both immortality and abiding love. Others say it is a rough simulacrum of the human face—without the forehead. Still others think of it as a yin-yang-like melding of the female and male principles: the top vaginal cleft and phallic bottom of an organ at once receiving and projecting the blood in that ongoing act of inner coition from and above which our consciousness continuously crests like mist above a flowing fountain.

Set on top of one of the cloth-draped metal surgical tables beside my left elbow was a clipboard with a form attached, some sort of official death notice. I was not supposed to see this, was not supposed to know anything about the donor except what Dr. Rosen has already told me. The form was turned away from me so that I had to read it upside down.

Just then, one of the OR attendants approached and took the clipboard in her hands, but it was too late. Certain details—Age: 27. Marital Status: Single. Height: 5'7". Eyes: Brown. Hair: Brown—had bled through, enough for my mind to start constructing an individual now around what had been better kept anonymous: a young woman who, not five hours ago, was fully immersed in her Saturday night; a someone, a very particular bundle of nerves, drives, and worries; of heartfelt expressions, expectant glances, and certain plans—the whole of that, with the sudden rupture of one tiny blood vessel, come crashing down into permanent darkness, and me there now, in the brash lights, standing over her.

"Heart-beating cadaver." Brain death. The sudden cessation of that lifelong to-and-fro between the heart and the brain, about which I'd recently been speaking with Bill and Carol and the other transplant candidates, but severed in this instance from the other end: the heart vigorous still, beating still, and yet entirely on its own, without any governance from the brain, without any "higher reason." Continuing to pump blood that—because the brain controls the breathing that oxygenates that blood—would have long ago gone blue-dark, eddied, and choked off the rest of her body's life if not for that respirator beside the table, her lungs rising and falling to the machine's whooshing clicks, her heart beating with no other compulsion but its own primordially instilled, deep cellular memory to do so.

I couldn't stop looking at it, a molten-lava red with fatty white striations across its surface, like vapor clouds around a hot, newly formed planet, the respirator's whooshes and clicks beginning to meld in my mind now with the clacking rhythm of wheels over train tracks: early April, 1980, only months since he "drowned in his bed."

I am sitting on a train, a midafternoon, suburban commuter train, going against the flow, going at an idle, off-peak hour toward downtown Chicago, trying, like some spent, limb-stranded blood cell, to get back to the heart, to the heart of something.

"Just get out of here," I remember being told by my mother, for whom I'd decided to delay my return to Houston in order to help her through her grieving. "You're driving me crazy."

I'm sitting on the train, watching the other passengers, my fellow breath-counters, oceanic outcasts, flyaway balloons, hating them all, one dark-suited businessman in particular, sitting in the seat across from me, whistling, idly unraveling his life's thread on the same repetitively tuneless jingle.

At the city's outer building edge, the train seems to pause a moment before beginning to draw us in a slow, steady submersion down past the different tenement windows: a nurse, still in her uniform, asleep on a bed; a man among swirling kitchen wallpaper patterns, leaning over an ironing board; someone reading in a lamplit chair, the window's curtain arms lifting outward in the train's wake.

From Chicago's Union Station I walk due east to Lake Michigan and then south along the shorefront toward the Shedd Aquarium. I have been there a number of times in the past few months, finding it restorative, somehow, just to watch the fish smoothly ply their walled and lighted days, wholly unaware, it seems, of their displacement, continually arriving, in warped carousel, at the edges of a wakeful, drifting repose: walleyed, tropical, floating among the coral branches like birds in some flip side of the sky; the nurse shark nosed into its sand bed; the lamplight fish, hoarding vision from darkness; the anemones, drowned stars, all their feelers out.

On this day, however, the aquarium is closed, a sign posted out front detailing upcoming renovations. There is still an hour or more of daylight left as I start back north along the lakefront, Chicago's spired concrete on my left shoulder, and on my right, just water and sky: the formed and the fluid so direly juxtaposed, only the harbor sailboat mast lines clinking in between, like a huge cocktail glass in the hand of a frivolous god.

I arrive at the front steps of the Art Institute, and there, I suppose in honor of my father, I abruptly decide to hop a cab instead and go to his favorite museum: the Museum of Science and Industry, near the campus of the University of Chicago, on the city's far South Side.

It is near closing when I arrive. I was there with him some years earlier. He proceeded straight to Life Sciences and Biology, to revisit, I suppose, with the earliest of his unlived lives, the would-be doctor who didn't consider himself smart enough. I rush inside, wind up clueless in a dank central stairwell that embraces a giant pendulum, its ceaseless whirling motion mirroring that of the earth, inducing in me the same dizzying pitch I now feel upon staring down into the donor's opened body.

Moving away from the railing, I keep my head down and climb to the next landing, where a series of thinly carved cadaver slices in pressed glass definitively mark the entrance to the Life Sciences wing. It is nearly deserted, and yet, late as the hour is, a small crowd of people still stands at the far back of the exhibit. I walk toward them, look past the assembled heads, find myself staring now at a darkened wall lit only by a row of fetuses.

They seem to drift within framed, backlit panes of formaldehyde, one after another, from pre-fetal zygote, to embryo, to developing fetus, the entire story of gestation all the way to the full-formed child. Moving in among the viewers, I begin to take in the frame-by-frame prospect of life assembling: stilled, fluid-bound swirls of cellular incipience, membranous infoldments, pinched-off proto-organs enclosed within the amnion—an ancient Greek word for a sacrificial plate to hold a victim's blood.

For the longest time I can't recall where I've encoun-tered these shapes before, and then it comes to me: illus-trations I'd seen of some of the earliest maps of the world ever drawn, ones dating as far back as 700 B.C., when the earth was conceived of as a flat disk of inchoate, globular landmasses floating within a blue, encircling sea called Oceanus.

"Yes, fine . . ." I could hear one of the organ harvesters saying in a clipped, purposeful monotone. "Let's have another look at that."

Someone was reaching now deep into the donor's body, sending a light blue tide of intestines riding up toward the beating heart, upon which Dr. Rosen and Dr. Slater had already set to work, they and the other teams' surgeons alongside them, cutting away at the connective tissues around their respective organs, readying them for imminent extraction.

"I can't use this!" the pancreas harvester announced. "It's way too fatty."

I watched a member of the liver team lift one end of their quarry now, a thick slab of midnight maroon that threw back the ceiling lights, the color alone suggesting why the liver was so long thought to be the blood's factory, fashioned there from a distillation of ingested food or "chyle," and then drawn upward into the heart's right chamber by a bellowslike expansion of the heart muscle, as though the heart itself were breathing: a separate animate entity, at once giving us, and living off of, our body's life.

Taking up his scalpel now, the liver harvester sliced away a tiny piece, placed it on a sterilized tray. An attendant rushed it out of the room for biopsy. Beneath the far wall clock hanging directly above the picnic cooler, Dr. Slater was on the telephone. Dr. Rosen moved away from the operating table, went over to the phone, took it in hand. The whole room was listening in.

"Yes, I see. Well, we're standing by."

"Doctor?"

It was the liver harvester.

"Yes, doctor?"

"Any idea when you might be cross-clamping?"

I'd remembered this term from Dr. Rosen's earlier rundown of the night's proceedings. "Cross-clamping" occurs at the moment just before the heart is extracted, the donor's aorta closed off with a clamp so that no more blood can flow into or out of the heart.

"No," Dr. Rosen said, "I can't really tell you at this moment."

"I am"—the liver harvester's gloved, blood-stained hands were flying up now in the lights—"so sick of this game. It's always the same thing with you heart guys."

"This," Dr. Rosen boomed, striding toward the table, "I can assure you, doctor, is no game."

Dr. Rosen began trying to explain about Columbia-Presbyterian's strict policy of minimizing the amount of ischemic time—the time in which the heart is without blood.

"Our recipient is just being prepped right now," he continued, "and he's a re-op. There's a good deal of built-up scar tissue from a former triple bypass."

The liver harvester seemed unimpressed. A well-built, square-jawed man about Dr. Rosen's age and at least a half-foot taller, he started moving away from the table. Dr. Slater, even slighter than Dr. Rosen, was still over by the phone. I started breathing heavily through my surgical mask, wondering if I was supposed to step in here, if this was to be the unlikely way in which I'd be asked to go beyond my observer status and "do something."

I saw the liver harvester gesturing toward the OR door. He and Dr. Rosen passed into the outer hallway, followed by Dr. Slater, and he by the other harvesters, the lot of them on the far side of the room's reinforced glass now, arguing, and me left alone there at the table with her, and her blossoming lungs, and her beating heart.

Chapter 16

A ND THEN MY MIND DID THIS. IT BEGAN—BY WAY of tempering such cold witness to what it itself is an inextricable part of—to make a story of what it was seeing, dim the room's too-bright lights, soften its brassy, sterilized edges, feel around for some other context in which a measure of awe and disbelief still attended such a raw exposure.

I found myself thinking back to the accounts I'd recently read at London's Wellcome Institute Library of the young William Harvey's anatomy classes at the University of Padua in the year 1599, and the many fracases that would erupt over the "exposed" subjects of his day. Back when the same body that lay open there before me that night was—like the physical world around it—yet to be fully encompassed by our minds, was still replete with its own expanses of spirit and serpent-filled terra incognita.

The anatomy amphitheater where the classes were held still stands. I traveled from London to Padua to see it, made the very journey that the twenty-year-old William Harvey did as a young medical student in 1598. It was a sort of pilgrimage for me back to the place and, in my mind, the relative moment in historical time when those two maps of our inner and outer worlds were, like that wall of lit fetuses, brought to their earliest completion.

I took the train to Italy. Harvey went on horseback. His likely route would have been through Paris, Troyes, Geneva, and Milan, a journey of at least two months, and a particularly arduous and perilous one, given travel conditions in the late sixteenth century. There is no record of his trip in

Harvey's letters or journals, but I was able to find a personal account of a similar journey made only forty years earlier by a young medical student from Switzerland named Felix Platter, whose later work on the human anatomy, De corporis humani structura, published in 1583, Harvey would cite frequently in the anatomical lecture notes he composed upon leaving Padua and assuming the chair of surgery and anatomy at London's St. Bartholomew's Hospital in 1604.

Platter was all of fifteen years old when, in October of 1552, he and a few companions left Basel for Montpellier in southern France, its university at that time, like those of Padua, Bologna, Vienna, and Paris, attracting students from all across Europe. In the month that it would take the young Platter and his companions to travel just over 400 miles, they came upon villages besieged by plague, had a number of close brushes with roving bands of thieves and cutthroats, and witnessed numerous public executions: of criminals; of local village Catholics by Protestants, and vice versa; and of anyone of either faith deemed a heretic by servants of the Inquisition, the guilty burned at the stake or flayed and dismembered, their body parts strung up in the branches of roadside olive trees as warnings to all passersby about the perils of contradicting Church dogma.

Harvey's friend and chronicler, John Aubrey, describes him as a diminutive figure, "not tall, but of the lowest stature, round faced, olivaster like wainscott in complexion," his eyes, "full of spirit," his long, wavy hair, "black as a Raven." And yet he is said to have had a particularly feisty temperament—"very cholerique," as Aubrey puts it, and also wore at all times a huge dagger at his left hip, forever worrying the pommel with his fingertips when he spoke, a defensive posture owing not merely to the perils of travel, it seems, but to the rough-and-tumble of day-to-day life in Padua.

Students from countries across Europe as well as the Middle and Far East, constituted nearly a third of Padua's population of 14,000, lending the small seaside town fifteen miles north of Venice a very cosmopolitan if somewhat riotous atmosphere. The student body was enrolled by nation, the French, the German, the English, and so on. Under the protective umbrella of nearby Venice, the capital of the larger Venetian Republic, then a bastion of freethinking and secular study, students were allowed to choose their own representatives, or *conciliarii* to the university's governing body. They could determine their curriculum and department chairs, and, as nations do from time to time, were also left to war with one another, either in arranged sports matches or all-out impromptu street brawls.

Even grand ceremonial events such as the appointment of a new rector were marked by a kind of sanctioned truculence, a curious custom known as *vestium laceratio* in which students would run down and tear the clothes off the new appointee, who then had to buy the items back at much-inflated prices. A statute was eventually passed in 1552 in an attempt to rein in the unruliness of such occasions, the sort of "too horrid and petulant mirth," as the statute worded it, that apparently did not stop at the anatomical amphitheater's door.

Erected in 1595, just three years prior to Harvey's arrival, it was designed by Girolamo Fabrizi d'Aquapendente, better known by the latinized form of his name, Fabricius, the university's Chair of Surgery and Anatomy. Carved entirely out of oak from the surrounding Paduan countryside, the theater consists of five concentric galleries, each a mere eighteen inches wide and pitched so steeply one above the next that, looking from the top gallery, I had only to lean slightly forward on the railing to see past the visitors in the lower tiers, directly down into the oval of the dissecting pit.

One floor beneath the pit is an even smaller oval stone chamber where, during Harvey's tenure, the demonstrations occurred, requiring even the first-tier attendees—a select group of university officials and physicians—to peer down upon the proceedings, the tabled cadaver appearing, in the many illustrations I'd seen, at once distanced and hyper-focused there, as though viewed through the wrong end of an oval spyglass: a splayed, fully formed fetus inside of a transparent egg.

The theater held up to 240 people, but demonstrations were regularly filled beyond capacity. Advertised for weeks in advance on printed flyers posted throughout town, anatomies at that time were public events and wildly popular, curious hybrids of instructional class and carnival sideshow, the theater's balconies draped in bright silk damask, Fabricius, along with the university rectors, all clad in flowing robes of purple and gold, hawkers moving among the crowd, selling oranges, bouquets of flowers, and incense to help mask the foul odors.

Lectures (which would often be spaced out over the course of a week) were held on average only four times a year and always during the winter months, subjects being both difficult to come by, and, in the absence of refrigeration or chemical preservatives, even harder to keep. They were, for the most part, the bodies of executed criminals, preferably "unknown and ignoble bodies, from distant regions," as one local law decreed, so that the subject could be dissected "without injury to neighbors and relatives."

It was also not uncommon at that time for medical students to rob from graveyards. Felix Platter mentions participating in a number of such expeditions while at Montpellier. Sometimes students would accost funerals mid-procession, so eager had people of the Renaissance become for a more prolonged and measured view of the body's inner vista than either the ravages of battle or of punitive displays allowed.

Italian audiences were said to be relatively sophisticated and well mannered. Still, admission being free, seating had to be strictly enforced. Members of the general public—those with no knowledge either of the medical sciences or of Latin, the language of the lectures—were consigned to the top tier. Guards were also posted at the theater's entrance to try to weed out those elements known to disrupt anatomical sessions in cities throughout Europe, especially when the subjects were female, people jostling for better views and shouting out lewd remarks the nearer the lecturer came to the area of the genitalia, some in the crowd rushing the table to touch the parts in question.

In the tiers below the top one would be a dizzying mix of distinguished attendees. Along with physicians and medical students, there were theologians, local magistrates, representatives of the papacy, members of the aristocracy in all their finery, some wearing carnival masks; and a number of notable scholars of all disciplines: philosophers, astronomers, physicists. Among the Paduan regulars were the famed astronomer and philosopher Giordano Bruno, a former professor at the university; Father Paolo Serpi, chief theologian of the free-spirited Venetian Republic as well as an accomplished physicist and anatomist, who is said both to have invented the thermometer and, along with Fabricius, discovered the tiny valves inside our veins that prevent the backward flow of blood. And then there was the university's most popular professor at the time, known to draw half of its 4,000 students to each of his lectures: the thirty-five-year-old Galileo Galilei.

The first phase of the lecture Fabricius devoted to the viscera, "the lower belly," as Harvey refers to it in his own *General Rules for an Anatomy*, "nasty yet recompensed by admirable variety." The last phase was reserved for "the divine banquet of the brain," and the middle for the thorax, known as the "middle belly" or "parlour," the various body parts dropped in buckets at the anatomist's feet as he made his way along. Set on small tables all around the cadaver would also be the opened carcasses of a variety of animals to which

Fabricius, an early proponent of comparative anatomy, would repeatedly refer in the course of his lecture.

Harvey, the *consiliarius* of the English nation and one of Fabricius's most diligent students, regularly got a spot in the amphitheater's first or second tier, especially coveted given the poor lighting. Anatomies were held in the early-morning hours, but, as the theater was situated well away from the building's outer windows, no daylight filtered in. Except for the faint flickerings of two candelabra positioned on either side of the top gallery, bearing three candles each, the only other light emanated from a ring of eight lamps held by students positioned around the rim of the dissecting pit itself, and whatever ambient glow still remained from the white mist that many of those in attendance at that time believed was released from a body when it was opened soon after death.

Mixed in with the cries from the upper gallery, and the verbal sparring among the student nations, there would be, as well, the requisite heated exchanges between Fabricius and the various scholars in the audience whose place it was, in the tradition of classical debate, to stand and pose objections and questions, setting off intense, rhetorically florid disputes over the true nature of our physical and spiritual makeup.

There came a sudden swell of voices as the harvest room doors opened again: a couple of surgeons coming back in to check on the donor's vital signs. I held my place, kept my head down, not wanting to catch anyone's eye, repeatedly reaching my right hand over to my left wrist for furtive pulse checks.

It was moving far too fast. Too much in sync with the heart I'd been staring at for so long, as though mine were living now through hers, bound to each one of its beats: the sharp, uptwisting quakes of systole, pumping the blood outward; and then the soft, near-simultaneous release as the tired blood returns, known as diastole, a term from Greek and Latin prosody referring to the lengthening of a normally short syllable.

From the earliest days of civilization, people pondered that simple two-beat motion, trying to divine from the outside the course of the blood within: all the physician-philosopher-astronomer-mathemetician-naturalist-humanist Everymen of antiquity, staring at everything from beating chick-embryo and insect hearts, to those of birds and mammals, and, in some instances, even the hearts of men.

Dissection of humans was rarely practiced in antiquity. The dead body was considered to be sacred. Still, at the liberal school of Alexandria in the third century B.C., the Greek physicians Herophilus and Erasistratus are believed to have performed not only dissections but occasional vivisections as well, slicing open the chests of condemned slaves, cutting into their hearts and the neighboring veins and arteries, trying to see—much like someone today barging into a photographer's working darkroom—the very picture they were destroying by coming in.

The idea of a self-contained, ever-recurring circulation

of the blood had been intuited long before that Paduan winter of 1599. According to a 3,000-year-old Chinese medical treatise, "The heart regulates all the blood of the body. . . . The blood current flows continuously in a circle and never stops."

A passage from Hippocrates, written in the fourth century B.C., states: "The vessels communicate with one another, and the blood flows from one into another. I do not know where the commencement is to be found, for in a circle you can find neither commencement nor end, but . . . these vessels . . . are like rivers that purl through the body and supply . . . [it] . . . with life; the heart and the vessels are perpetually moving, and we may compare the movement of the blood with courses of rivers returning to their sources, after a passage through numerous channels."

Still, it was the tracing of those inner bodily channels and the cycle, nearly fugue-like in its intricacy, that the blood makes through them that would prove to be so elusive, primarily because no one could find a way to slow it all down, could figure out how or where in the heart's and blood's ongoing motion to intercede without disrupting that perfect simultaneity.

I walked through a human heart once, took a train from New York to Philadelphia. They have a giant "Walk-Through Heart" there, standing alone in a dimly lit corner of the Franklin Institute's Life Science wing. About fifteen feet high and nearly that many in diameter, the heart is roughly the size of your average Manhattan studio apartment and just as noisy, loud recorded lub-dubs filling the Institute's outer rooms and halls.

To enter the heart, you have to walk up a short flight of stairs that lead into the superior vena cava and then pass— just as our spent, returning venous blood does—into the heart's upper right atrium. There were a number of children inside, knocking about the chamber, dangling, like animate blood clots, from the edges of the tricuspid valve. I seemed to have been the only adult in there with them, the others remaining on the heart's outskirts, waiting impatiently for their kids to complete the circuit as though it were a carnival kiddy ride.

My initial impulse was to turn around and flee, my chest tightening, my breath getting shallow, my mind conjuring a series of mortifying newspaper headlines: "Man Has Heart Attack Inside Heart!" "Man-Sized Clot Lodges in Heart!" But as there appeared to be more kids coming in behind me than I could hear milling around up ahead, I decided to keep going, pushing my way down through the tricuspid valve into the right ventricle, which, to my great relief, I found empty.

It seemed to me as good a place as any to gather myself: the precise juncture in the blood's circuit at which the human imagination snagged for thousands of years, unable to construe the idea of a "lesser" circulation within the larger one. Unable to make that same supple, whale-like leap that our blood does away from the heart, up into the lungs' outer air, and then splashing back down again:

a brief circulatory aside that is at once the result and the ongoing recapitulation of that very climb life itself long ago made up out of the sea.

More voices now. I looked up. One of the harvest room attendants was stepping out into the hallway. Another attendant checked the respirator. I held to my spot, fixing my gaze, once again, on the beating heart, trying to imagine what the young William Harvey and all the others in that Paduan amphitheater 400 years ago would have made of this otherworldly prospect there before me: the body's dark, inwardly purring projection room opened wide now and working still, but with no light behind the eyes, that light which, in the late sixteenth century, was believed to be man-ufactured in the heart.

The heart was, for much of its history, thought of in a far more mechanical light than it has come to be even now: as a kind of catalytic chamber of the spirit, igniting, in turn, the body's larger house of spirits, all those murky vapors like melancholy that were long thought to infuse and exhaust us. In fact, Galen's second-century-A.D. model of the circulation—which, like his contemporary Ptolemy's Earth-centered universe, was still being taught in 1599—sounds to a modern ear like some early biological proto-type of a two-stroke internal-combustion engine.

The liver-enriched blood, having already passed through the veins to nourish all the organs, was thought to arrive back at where I was resting in the right ventricle that day as a kind of sooty residue, which was then believed to leave the body through the lungs as exhaust during exhalation. A small fraction of the sooty blood, however, was thought to seep over sideways to the left ventricle through tiny pores, or *foramina* as the Italians called them, in the septum, the dividing wall between the heart's left and right chambers.

Once inside the left ventricle, this sooty blood was re-mixed with and sparked by air or *pneuma* which the heart drew into itself, bellowslike, from the lungs during inhalation, thus creating a kind of animating spirit known as "life pneuma." The downward thrust of the next exhalation then forced this inspired admixture from the left ventricle out into the arteries to give life, literally, to the body's different organs.

That such an ancient model of the human anatomy should sound so mechanical is not the paradox it would seem. The word "mechanical" derives from the Greek *mekhanikos*, from *mekhane*, contrivance; or *mekhos*: as in "means to an end." Galen's conception of the body, along with Ptolemy's of the universe, with its tightly ascending rings of planets, stars, and upper galleries of cherubim and seraphim; or those earliest maps I'd seen of the Earth—were naturally more contrived, abstract, two-dimensional constructs, proceeding as they all did from a necessarily narrower scope of available information and awareness about the separate fields of focus they were meant to portray.

Anyone looking now at representations of the Ptolemeic heavens, or at depictions of the Galenic anatomy right up through the late Middle Ages, such as those in the fifteenth-century science textbook *Marguerita Philosoph-ica*—cartoonish squatting figures, arms and legs akimbo, the exposed, Latin-labeled inner organs arrayed like the quadrants of a butcher shop's cow poster, recognizes them to be approximate, symbolic representations of reality rather than rigorous attempts at verisimilitude.

By the winter of 1599, however, those separate fields of focus were all coming into far clearer and fuller view. Galileo was regularly corresponding at that time with the German astronomer Johannes Kepler about Nicolaus Co-pernicus's new heliocentric model of the universe. The Earth had already been circumnavigated three times (most recently in 1580 by Harvey's countryman, Sir Francis Drake). The inner depths and shades of the human form, mean-while, were being fully plumbed and lifted into the light of day: in the public anatomies conducted throughout Europe, most famously by the so-called Father of Anat-omy, Andreas Vesalius; and, conceptually, in the paintings and etchings of the Renaissance masters, among them Jan Stephan van Calcar, the Dutch painter and student of Titian who composed the vivid anatomical illustrations for Vesalius's revolutionary anatomy textbook.

A number of those sitting in Fabricius's amphithea-ter that winter of 1599—Giordano Bruno, Paolo Serpi, Galileo, Harvey, even Fabricius himself—knew there were no "foramina" in the heart's septum. Leonardo da Vinci's intricately detailed late-fifteenth-century depictions of the numerous hearts he'd dissected had already hinted at the impermeability of the septum, a fact that Vesalius, himself a former Anatomy Chair at Padua, was to confirm repeatedly in his own demonstrations and writings. But everyone in that amphitheater knew, as well, the perils of voicing such an opinion too strongly.

As openly disputatious as those sessions could become, things quieted down quickly when the lecturer's pointer arrived at "the parlour's" main occupant. Cut into the amphitheater's outer walls at that time were a number of "Judas holes" through which spies for the Inquisition would observe the proceedings, listening for the slightest contradiction of Galenic doctrine, especially when it came to the subject of the heart.

Vesalius would eventually lose his chair at the university over his increasing insistence on the impermeability of the septum, and was forced to flee Padua. Paolo Serpi had already been imprisoned once for his provocative writings about the movement of the blood in the veins and the suggestion of an inner *circulatio*, one that seemed to him to mirror that of the planets themselves as he'd often heard it described by his close friend Galileo.

Circulatio was, in fact, fast becoming the topic of conversation among the cognoscenti of late-sixteenth-century Padua. Realdo Columbo of Pisa, Fabricius's immediate predecessor as Anatomy Chair and a student of Vesalius, had published a book in 1559, the year of his death, in which he described the "lesser" or pulmonary circulation, a discovery that research now shows a particularly ingenious Arab physician named Ibn al-Nafis had already hit upon as early as the thirteenth century.

Columbo's former student at Pisa, meanwhile, Andreas Cesalpino, philosopher, botanist, court physician to Pope Clement VIII, and a friend of the aging Michelangelo, published his own anatomy text in that same year of 1559, in which he seems to have coined the very word *circulation*, referring to "this circulation of the blood across the lungs," and then going on to write a sentence that, although he was unable to provide the clearcut evidence to support it, suggests he was aware as well of the concept of a larger circulatory system: "We see the blood to be induced to the heart through the veins and then distributed to the whole body through the arteries."

Both Columbo and Cesalpino, however, were close friends of the papacy. Vesalius was not so fortunate. A few friends in high Venetian circles would enable him to avoid the wrath of Inquisitors for a time. But in 1562, while serving as personal physician to Philip II of Spain, he inadvertently offered his enemies their long-sought pretext for permanently silencing him.

The circumstances are somewhat vague, but it seems he began to perform an autopsy on a subject who proved to be still alive. The victim's family appealed to the Holy Court at Madrid. Vesalius was sentenced to burn at the stake. King Philip's last-minute intervention did get the sentence commuted to a "voyage of atonement to Jerusalem" a long journey by sea, and one that, judging from the accounts I'd found at the Wellcome Institute Library in the journals of one Italian sea traveler in the year 1574, must have been its own form of torture:

The dangers and trials I suffered were boundless and terrifying. True, the ships are large and strong, but they were so overcrowded with passengers, merchants, and merchandise that there is little room left for anyone to move around in. The ordinary passengers on deck are obliged to stand all day in the broiling sun and sleep in the open in the cold of the night. On the other hand, the cabins provided for the nobility and the rich are so low and narrow that a man can do little but crawl in. . . . The meat and fish are so salty that the suffering resulting from their consumption is indescribable. . . . Another difficulty, and the greatest, is the lack of water. Throughout the journey, the daily ration distributed is so putrid and evil-smelling . . . passengers are obliged to hold a piece of cloth before their mouth to filter the putrefaction. The liquid is distributed once a day and many do not receive even this, . . . many die of thirst. Still another trial stems from the various diseases contracted by the passengers. . . . Often a large part of the passengers die, . . . sometimes three to four hundred on a single ship, and it is heartbreaking to see the unfortunate bodies every day as they are thrown out to sea.

In 1564, at the age of fifty, Vesalius would leave Jerusalem, bound for Venice, having received a sudden invitation from the Venetian Senate to resume his Anatomy Chair at Padua. But the ship he boarded was lost in a storm. A fisherman who is said to have found Vesalius washed up on the shores of Cyprus ran off, thinking him a victim of the plague. A nearby village goldsmith did eventually take him into his care, but would be unable to save him.

I often found myself, in the course of reading historical accounts of that period, wondering what it was about those "foramina," aside from the accrued weight of centuries of tradition, that had the authorities of the day so direly clinging to them and the larger Galenic model. And then it occurred to me that the particular genius of that construct, especially in terms of pre-Renaissance conceptions of the human being and our place here on Earth, was that it so clearly incorporated the most primal and essential aspect of all human creation mythology: the first life-giving breath, the spirit of a divine Creator.

The word *pneuma* means not merely air but the soul or vital spirit: blast of wind, breath, divine inspiration. Ancient Chinese Taoists held that in the beginning there were nine breaths that eventually coalesced to constitute physical space, the word *ch'i* referring to the intermediate space between Heaven and Earth in which we humans live "like fish in the water." Ancient Hindu teachings referred to that same intermediate realm as *vayu*, the wind and breath of life, and *sutra* as the thread of the universal spirit that ties the two worlds together. The Muslim word *er-ru* and the Hebrew *ruah* both invoke God's formative, life-giving breath or spirit. In the Book of Genesis, Jehovah imparts into Adam's nostrils the *ruah*, or original breath of life.

Spiritus or "life principle" even informs some of the earliest known scientific theories of embryology, in which the combined "seed" of the man and the woman was thought to have its own inherent, life-impelling breath. I found an extraordinary passage at the library one afternoon, believed to have been written by the physician Polybus, son-in-law of Hippocrates, in a treatise titled *The Nature of the Child*. In it he describes a six-day-old human embryo that he happened to obtain one day through somewhat odd circumstances on the island of Cos in the fourth century B.C.

It was in the following way that I came to see a six-day-old embryo. A kinswoman of mine owned a very valuable danseuse, whom she employed as a prostitute. It was important that this girl should not become pregnant and thereby lose her value. Now this girl heard the sort of thing women say to each other—that when a woman is going to conceive, the seed remains inside her and does not fall out. She digested this information, and kept a watch. One day she noticed that the seed had not come out again. She told her mistress, and the story came to me. When I heard it, I told her to jump up and down, touching her buttocks with her heels at each leap. After she had done this no more than seven times, there was a noise, the seed fell out on the ground, and the girl looked at it in great surprise.

It looked like this: it was as though someone had removed the shell from a raw egg, so that the fluid inside showed through the inner membrane— a reasonably good description of its appearance. It was round, and red, and within the membrane could be seen thick white fibers, surrounded by a thick red serum; while on the outer surface of the membrane were clots of blood. In the middle of the membrane was a small projection: it looked to me like an umbilicus, and I considered that it was through this that the embryo first breathed in and out. From it the membrane stretched all around the seed.

Now, the formation of each of these parts occurs through respiration—that is to say, they become filled with air and separate, according to their various affinities. Suppose you were to tie a bladder onto the end of a pipe, and insert through the pipe earth, sand and fine filings of lead. Now pour in water, and blow through the pipe. First of all the ingredients will be thoroughly mixed up with the water, but after you have blown for a time, the lead will move toward the lead, the sand toward the sand, and the earth toward the earth. Now allow the ingredients to dry out and examine them by cutting around the bladder: you will find that like ingredients have gone to join like. Now the seed, or rather the flesh, is separated into members by precisely the same process, with like going to like.

To the late-sixteenth-century imagination, the notion of a self-contained, continuously recurring circulation of the blood seemed to evict the pneuma from our system, to sever that original, connective life-thread between ourselves and the divine. In effect, it was a system that failed, as my longtime childhood tormentress, Sister Mary Margaret, would have put it, "to leave room for the Holy Ghost."

Still, the human imagination has always proven adept at resituating the soul, at finding some other place within us for the spiritual. As Harvey himself would later insist upon publication of De Motu Cordis, his deepest longing—and it is the very one that a devout Galileo would soon find himself expressing to his inquisitors—was only to further prove, via the evident complexity and wonder of what he'd discovered, the truly divine scope of God's genius.

The most unsettling aspect of circulatio—be it that of the outward universe, or the one within our own body's inner microcosm—was its connotations of speed, simultaneity, and flux. Galileo's telescopic verities didn't—as it is often assumed now—land him in prison because they unseated the Earth from its exalted place at the center of the universe. The Earth in Ptolemy's schema, and certainly in the mindset of the late Middle Ages, lay, much like that tabled cadaver there in Fabricius's amphitheater, both fixed and as yet unfound at the bottommost rung, the one farthest from heaven and a tier above hell.

Galileo and his new spyglass were about to make that fixed scheme take flight, to introduce us into the larger unknowable mix: a whirling sphere among whirling spheres,

and we humans now—in the words of the frightened Flo-
rentine ambassador who'd been sent along to Rome with
Galileo to help him face his inquisitors, "flying around in
circles with [it], like ants on a balloon." Harvey's circulation
was poised to do the same with us and our psyches: to set us
off forever on our own inwardly roiling, circulatory seas. All
he needed was his own way of seizing and encompassing
the system long enough to see.

Then one afternoon, while walking home from the
amphitheater, Harvey passed through the town's tightly
wound, cobbled streets into the central market, and noticed
a fish still flopping atop a monger's cart. An avid comparative anatomist in his own right, Harvey spent hours in his
room in Padua that winter of 1599, just as he'd done as a student back at Cambridge, peering into any creature he could
get his hands on. He'd cut open flies, earthworms, wasps,
roaches, the smallest life-forms he could find, trying to
answer the very question I'd asked Neal Epstein in his office
that first day about whether there were any living things that
didn't posses a heart, trying, with the new pocket lens he'd
been given by Galileo, to fathom the frenzied pulsations of
the simplest tubelike structures.

He cut into live mice, frogs, pigeons, dogs. He even
once got hold of a stillborn baby and, as though trying to
test Polybus's breath-in-the-seed theory, blew through the
umbilicus to see if there were any corresponding movements in the child. Harvey's chambers soon came to be
strewn with the cast-off carcasses of his curiosity, a scene
he would no doubt have recognized from another passage

in the writings of Hippocrates, the very passage, in fact,
that Harvey's countryman Robert Burton would later cite
in the introduction to his classic *Anatomy of Melancholy*,
published in 1621, seven years before *De Motu Cordis*.

Explaining why he chose as his pseudonym the name
"Democritus"—fourth-century-B.C. philosopher, mathema-
tician, student of natural history, and one of the original
formulators of the atomist theory—Burton writes:

I shroud myself under his name . . . , for that reason
and only respect which Hippocrates relates . . . , how
coming to visit him one day, he found Democritus
in his garden at Abdera, in the suburbs, under a
shady bower, with a book on his knees, busy at his
study, sometimes writing, sometimes walking. The
subject of his book was melancholy and madness,
[and] about him lay the carcasses of many several
beasts newly by him cut up and anatomized, not that
he did condemn God's creatures, as he told Hippocrates, but to the intent he might better cure it
in himself, and by his writings and observations
teach others how to prevent and avoid it.

Harvey hurried the fish back up to his room. Setting it
on a table before him, he held it down, feeling against his
palm that firm, full-length press of its body as he cut the
fish open. He sat staring at its sluggishly beating heart.
Time slowed, all the clamor of his day at the anatomy lecture fading away with the world outside his window. He

reached in and massaged the heart. Taking it between his thumb and forefinger, he felt that seemingly simple, two-beat motion: the rigorous blow of systole, and then diastole's sudden, flaccid calm.

He next took up a short length of string, reached in, and tied off the aorta just above the heart. He watched as successive waves of pumped blood backed up into the heart, filling it to bursting, the section of aorta on the far side of the ligature flattening for lack of blood. He next reversed the exercise, tying off the main vein leading into the heart, watching as the blood backed up in it and how the heart flattened and the aorta with it.

Over the coming months, Harvey would repeat these experiments and observations countless times with whatever fish he could find. He became obsessed, as I would in those aimless months after my father's death, with staring at fish. Their hearts, he soon discovered, not only beat at a more plodding and measurable pace than ours; they beat for a good length of time both after they had been cut out, and cut into many pieces, the muscle shards themselves continuing to pulsate before him, pulsating with no other signal or governance but their own primordially instilled, deep cellular memory to do so.

"When in many dissections, carried out as opportunity offered upon living animals," he began to write one night in the notebook in which he'd been gathering observations and ideas for De Motu Cordis, "I first addressed my mind to seeing how I could discover the functions and offices of the heart's movement through the use of my own eyes

instead of through books and the writings of others, I kept
finding the matter so truly hard and beset with difficulties
that I all but thought that the heart's movements had been
understood by God alone. This was because of the rapid-
ity of the movement, which in many animals remained visible for but the wink of an eye or the length of a lightning
flash, the movements now diverse, and now inextricably
mixed. Hence my mind was all at sea. . . .

"But these features," he continued, "are more obvious
in the hearts of cold animals, such as toads, serpents, frogs,
and fishes. First, then, in fishes, which provide clear confir-
mation of the truth, for everything takes relatively longer
and is much more distinct."

He paused a moment, took up his pen again, stared
out his window toward the banks of the River Po, then
pulled his table lamp closer:

There are, as it were, at one time two motions. Unless you deny the evidence of your own eyes, you
must acknowledge that the blood returns to the
heart, that it has a movement, as it were, in a circle.
We have as much right to call this movement of the
blood circular as Aristotle had to say that the air and
rain emulate the circular movement of the heavenly
bodies. The moist earth, he wrote, is warmed by the
sun and gives off vapors which condense as they are
carried up aloft and in their condensed form fall
again as rain and remoisten the earth, so producing
successions of fresh life from it.

It may very well happen thus in the body with the movement of the blood. All these happenings are dependent upon the pulsative movement of the heart. This organ deserves to be styled the sun of our microcosm just as much as the sun deserves to be styled the heart of the world.

Harvey would wait nearly thirty years to publish these thoughts in his *Exercitatio Anatomica de Motu Cordis*. The year was 1628. In the thirty years since he had first traveled to Padua, Fabricius would publish *On the Formation of the Fetus*; Galileo his *Sidereal Messenger* (describing the moons of Jupiter as seen through his newly perfected spyglass); Johannes Kepler, his laws of motion in the *Harmony of the World*; William Gilbert, *On the Magnet*; Robert Burton, his *Anatomy of Melancholy*; and Francis Bacon, *The Advancement of Learning*. The heart's figurative map, meanwhile, would be forever broadened and enhanced with the publications of Miguel de Cervantes's *Don Quixote* and the major works of William Shakespeare, including *Hamlet*, *King Lear*, *Macbeth*, and *The Tempest*.

In that same thirty-year span, Giordano Bruno was lured into an ambush by servants of the Holy Inquisition and hustled off to Rome, where he was burned at the stake. Paolo Serpi, who had once again run afoul of papal authorities, was attacked late one night on the streets of Venice by assassins believed to have been hired by the Pope himself. His friend Fabricius had to be summoned from Padua to remove the stiletto that had been stuck in Serpi's upper jaw. In 1622, Galileo was brought before the papal court in Rome and placed under house arrest.

In 1642, Harvey's London lodgings near St. Bartholomew's Hospital were plundered by Oliver Cromwell's Parliamentary soldiers during the Civil War. The original drafts of Harvey's books and all of his papers were stolen, including extensive observations on the anatomical structure of animals and insects. Harvey himself, for all of his achievements and accolades, would be heckled throughout the rest of his career as "the circulator"—a man of "circular reasoning."

One rival physician even went about claiming—as though trying to stay the world in an age of unfinished maps with dragon-filled *terrae incognitae*—that at a recent public anatomy he'd attended, the anatomist found coiled up within the heart's left chamber "a worme or serpent" that would have been flushed out had blood really flowed there.

I remember getting more or less flushed out of the Walk-Through Heart that day. A group of kids came rushing down through the tricuspid into the right ventricle, forcing me up through the skylight-like pulmonary valve, out toward the lungs. Midway along the pulmonary artery— I suppose in deference to high-strung claustrophobics like myself—the artery's upper half or roof was cut away, so that I suddenly found myself standing back out in the open air again, perched on what was essentially an arterial balcony just below the heart-room's ceiling.

I paused there for a moment, looking back down on the uneven flow of museumgoers below, some of the parents snapping pictures of the children beside me, waving their hands and screaming, "We're on top of the world. We're on top of the world!" Then I turned, put my head down, and completed the rest of the blood's circuit in one blind rush: the quick oxygen grab in the lungs—wherein I was treated to the requisite whirring, wind-tunnel effects; the subsequent return to the left atrium via the pulmonary vein, and then the final drop down through the mitral valve into the main pumping chamber of the left ventricle, which propels the revitalized blood up through the aorta and out to the body again.

I felt more relieved upon exiting than inspired, going over to sit for a while on a small wooden bench by a wall of heart-related exhibits. There were a number of mementos and poems from hospital cardiac patients. One was by a seven-year-old boy named Louis:

> If you didn't have a heart,
> you wouldn't be able to breathe.
> If you didn't breathe
> you wouldn't be living.
> If you wouldn't be living
> you would be dead.
> If you were dead,
> you would be underground.
> If you were underground,
> you'd be in a coffin,

under a gravestone,
in a graveyard.

Alongside the case of heart poems, there was a squeez-
able hand lever meant to simulate the force of each heartbeat; and there were a number of exhibits to test people's
knowledge of the heart's workings, one challenging visitors to trace the very circulatory route that I had just made.
In the entire time I sat there, at least fifteen minutes in the
intermittent screams and lub-dubs, not one person was
able to do it correctly.

It's odd about the really great discoveries—the circulation, both of the blood and of the planets around their
respective, pulsating centers; Kepler's laws of motion; New-
ton's late-seventeenth-century work on universal gravitation;
and so on—how readily dismissed they are not only before
but after their acceptance. Concepts so seemingly beyond
comprehension, prior to their discovery, and then so in-
controvertibly correct once revealed, that we more or less
dismiss them as obvious. Somehow the great discoveries
become the equivalent of those inscrutable tinkerings the
TV repairman makes, readjusting and restoring what
seems to us like the same old picture, that flickering surface
story we all live by.

Chapter 17

O UTSIDE THE HARVEST-ROOM'S WINDOWS, HANDS were still flying, separate cells of dispute having arisen now around the core argument, which, it suddenly occurred to me, Dr. Rosen may have been purposefully drawing out for the benefit of the transplant surgeons back at Columbia-Presbyterian, trying to grab them as much time as possible before he finally cross-clamped and extracted the heart.

Time seemed to me a remote memory by then, something over there on a wall above a picnic cooler. There was only the whoosh and click of the respirator, and the presence of the donor—this by-now-animate auto-body parts shop—opened there before me. I felt as though I were in a room far beneath the earth, beneath daylight and thoughts and beliefs, beneath all the workings of the brain that both complicate and complete us.

All of that was dead. This was the elemental limbo of organs and of organ-letting, an ultimately brief abeyance in whatever ritual of death and mourning that young woman's family wanted for her. An otherworldly grace period that they had granted us, the harvesters, in order that someone else back in the waking world, if we could ever return there, might continue to ply it.

"This might," I could hear Dr. Rosen saying now, as the operating-room doors swung open again, "be a discussion for another time, doctor."

He went directly to the wall phone, Slater following closely behind. The other harvesters began to reposition themselves about the table.

I did another pulse check, the dizziness I'd felt upon first reporting to the table having returned.

"Hold this," a voice said.

Dr. Slater, handing me a clamp, trying, I realized later, to keep me in the game, one that, without my having even noticed the go-ahead signal, had now commenced in earnest.

"Clamp?"

I handed it back to Dr. Slater. He reached in, awaited Dr. Rosen's order.

Systole . . . diastole . . . systole . . .

"Okay, cross-clamp!"

Dr. Rosen injected the cardioplegia fluid, having timed the injection, he would explain to me later in the course of the ride back to Columbia-Presbyterian, at a point of posi-tive energy, the blood having just been pumped out of the heart so that it would take in more readily the very liquid designed to arrest it. The sudden infusion of potassium surrounded the heart-muscle's outer walls, preventing its cells from releasing the potassium that they naturally do, in exchange for sodium, with each contraction, thereby seizing the heart, literally, mid-beat.

Dr. Rosen now set about severing the heart from its many moorings—the venae cavae (superior and inferior); the pulmonary artery and veins; the coronary artery; the aorta. Dr. Slater, meanwhile, began taking up the iced bags of saline solution, ripping off the tops, pouring them, one after the next, over the now-stilled heart in order to pre-serve its outer tissue, bright torrents of bloody saltwater cascading off the table, splashing onto the tiled floor, running over and around our shoes.

In a moment, there was a freed heart cupped in Dr. Rosen's hands, intently cradled as one would a handful of water. He walked it over to a metal bowl set on a small table beside the Gott cooler. Dr. Slater and I followed him, the first time I'd moved my legs in over two hours. I watched him set the heart down in the bowl. He and Slater bent directly over it, poking and prying, turning it all angles. Slater caught me staring in, wide-eyed.

"Just want to be sure we have everything," he said.

He took up the heart, slipped it into a clear plastic bag, filled the bag with more ice-cold saline, tied off the top, then double-bagged it and placed it in the cooler. Across the room, the other harvesters were working feverishly, their respective organs having lost their life-source now, the operative time that much shorter.

"Thank you, everyone," Dr. Rosen called out as we wheeled our way toward the OR doors.

"Yeah," the liver harvester said, lifting only his eyes. "Thank you."

Chapter

18

THE RAIN HAD STOPPED BY THE TIME WE GOT back out to the New Jersey Turnpike, nearly empty on an early Sunday morning, the sun just coming up behind the Meadowlands' smokestack-studded swamps, our sirens blaring, the picnic cooler tucked away in the trunk.

We were doing at least ninety miles an hour, and yet, in our collective urgency to get our package home, it somehow seemed we were hardly moving, the road distending into distance, that iced heart's memory, I kept worrying, fading with each passing second.

I couldn't help asking Dr. Rosen about the tussle back in the operating room, wondering if there were any medical reasons for the liver surgeon's blow-up.

"Not really," he said, a wry smile on his face. "With brain death, certain changes do occur in the body that require changes in the medications we give to the donor. After enough time, some of those medications might do damage to a liver, but that really didn't come into play in this case. What you were seeing back there was just your typical type-A surgeon personality on full display."

"It's also sort of an emotional strain, performing surgery in such circumstances," Dr. Slater said, his eyes fixed straight ahead. "Even though the entire procedure is kept so separate from the donor's family. But there was this one time. Dr. Rosen and I had to fly out to Michigan to harvest a pediatric heart. It belonged to a four-year-old boy who'd died in his home of a blow to the head. When we got to the hospital for the harvesting, the parents were right there. They saw their son all the way to the OR doors, as though he were going in for a routine surgery. It was just crushing."

I sat there, staring off into the early-morning light, thinking about the heart in our trunk, its lifetime of cellular memories waiting now to be imparted to someone else. I thought, once again, about all the excised hearts of centuries past, the still-throbbing heart sacrifices and the already dead ones, cut out expressly for their own "heart burials," separate funerals for the life that lived within us, our first confidant: the hearts of royalty and nobility, of knights of the Crusades, and of martyrs and saints, each one mummified, licoriced, nitred, dipped, encased, jarred, cupped, entombed, whole churches built around the gilded reliquaries holding them. Each one given its own vain shove toward immortality.

I remember one afternoon, in the course of my stay in London, paying a visit to the recently restored nineteenth-century operating theater on the site of London's old St. Thomas Hospital, originally founded in 1106. The operating theater, tucked high up in the attic of what was formerly the St. Thomas parish church, is remarkably similar to Fabricius's sixteenth-century design, right down to the tightly wound outer spiral staircase leading up to the same five steeply pitched, concentric wooden viewing galleries. From there, other doctors and students would observe the surgeons, clad in their purple frock coats and carpenter-like aprons, performing surgery (such as it was at the time—amputations mostly), the writhing, ill-anesthetized patient held by assistants to a narrow wooden table set above a box of sawdust.

Upon exiting the theater that day, I made my way across St. Thomas' Street toward the entrance of Guy's Hospital, whose medical school, according to my guide-book, featured an anatomical library with one of the most extensive collections of specimens in the world.

The library, as the guidebook made very clear, is closed to the general public, but I decided to see how near to it I could get. Passing through Guy's metal front gates into a building-rimmed courtyard, I spotted a security guard posted at the entrance to what appeared to be a dormitory, and asked him directions.

It was a warm late-spring afternoon, the sight of med-ical students sitting out in the small tree-shaded closes between buildings setting off a memory of a passage I'd recently read from the letter of an early-seventeenth-century medical student at Christ's College, Cambridge, where Harvey had studied just before leaving for Padua.

"Going on Wednesday from Jesus College pensionary with Dr. Ward to his College . . . ," the letter, dated April 16, 1631, begins, "and espying a garden open I entered and saw there a hideous sight of the skull and all other bones of a man with ligaments and tendons hanging and drying in the sun by strings upon trees. I asked what it meant. They told me it was the peddler they anatomised this Lent and that when his bones were dry they were to be set together againe as they did naturally and so reserved in a chest or coffin for their use who desired such an inspection."

Finding no guard at the front security desk of the building that was said to house the anatomical library,

I hurried past a STUDENTS MUST SHOW I.D. sign, then ducked through a glass door that let onto a stairwell. One flight up I found myself in a long hallway of closed, opaque glass doors with numbers on them. A student directed me to the far end of the hallway, where, at the top of a small set of unlit stairs, there was yet another door, far too sequestered, it seemed, to live up to its stenciled billing: ANATOMICAL LIBRARY.

I took hold of the doorknob, turned it slowly, listening for the click, then eased my way inside, nearly colliding as I did so with a short, bow-tied gentleman in a tweed coat, wheeling a human skeleton.

"Yes?" he said. "Can I help you?" the dangling bones clacking alongside him.

We were in some kind of anteroom lined with filing cabinets and shelves. A wooden swivel chair was set before a wall-length desk covered in papers. A small black nameplate amid the confusion confirmed for me that I'd entered the chief librarian's office. He stood there by his still-swinging skeleton, patiently absorbing another of my rambling attempts to explain myself: the fact that I was in London doing research for a book about the heart, what sort of book I wasn't entirely sure, and clearly it would have been better of me to have made a formal request in advance, but I'd only just read of the library and happened to be . . .

"Well, then," he interjected, "what good are rules if we can't occasionally bend them, yes?"

He took up a pad of paper, wrote down my name and, beneath it, the words "some sort of book about the heart."

He then escorted me through a door at the back of his office that opened onto a gangplank-like structure with railings on either side, leading up to what appeared to be the top of a huge wooden cask, at least five stories high.

Proceeding through a dimly lit passageway, we emerged onto a narrow, semicircular balcony, bound by a brass railing, the dark wooden walls all around us covered, as were those of every tier below, with lit display cases of human body parts, preserved in the same sealed panes of formaldehyde that I'd seen years earlier at Chicago's Museum of Science and Industry.

"Now the hearts would be just there, around the corner to the right," the librarian told me, pointing one floor down to where the main circular body of the library gave on to a whole other wing.

"But indeed," he said, turning to leave, "feel free to wander about as you please."

Staring down upon the lower tiers, I could see, here and there, groups of students crammed up against the railings, listening and writing notes as professors lectured on whatever body part they'd assembled before, the entire human anatomy neatly parceled out, numbered and catalogued. I passed hands and feet that had been beset by all manner of afflictions—gangrene, elephantitus, guinea worm—parts either ghostly yellow or blackened from necrosis. There was a wall of kidneys, livers, and lungs, and one of sliced and pressed faces in profile.

Turning the corner to the wing where the hearts were kept, I saw down on the first floor a room of glass cases containing the horribly deformed bodies of various birth defect victims. Just across from there, at a bank of flashing computer screens, two young medical students were flirting alongside a lit display case of severed heads.

The hearts took up an entire corner of the wing. Hundreds of them. All those heart diseases I'd read about in medical dictionaries over the years set out there before me now in three, sickeningly faded dimensions: thickened, impeded hearts like my father's; greatly distended, balloon-like hearts; tiger and tabby-cat, and hairy, and half-formed fetal hearts. On and on they went, each with its own number at the bottom of the pane, a number which, it soon dawned on me, corresponded with the pages of a catalogue set on a small table along the brass railing behind me, each page containing a biographical sketch of that particular heart's former owner.

VERNON QUICK. Born: 1917. Died: 1979. Married. Father of three. Profession: Dockworker. Out of work for the last ten years of his life. Heavy alcohol intake. Cause of death: dilated cardiomyopathy, sometimes referred to as "Beer Heart."

MARY SWANSON. Born: 1934. Died: 1963. Single. Profession: Cabaret singer/Comedian. Cause of Death: Ruptured Aorta.

I continued thumbing through the pages and then moving back to the display case to inspect the very heart I was reading about, its autobiography in a sense, or the story of the life which that heart had authored, all those

249

millions of fly-wing beats come to rest here now within a
lighted wall of fellow hearts, as near to immortality, perhaps, as this life allows.

Not the end yet for our heart, I kept thinking, as we sped
north along the New Jersey Turnpike. Still more beats in it.
Still more light to give to someone.

A few weeks before the harvest, I had attended the annual party at Columbia-Presbyterian for transplant recipients and their families. It was held in a makeshift ballroom
on the ground floor of the Milstein Hospital building.
Hearts were everywhere that night: heart-shaped tablecloths
and streamers, and centerpieces that looked like miniature
hot-air balloons, the bright red, foil-wrapped gondoliers
below sprouting tiny heart balloons of their own. Most of
the guests had honored the party tradition of wearing red
and white. The band played only heart numbers: "You Gotta
Have Heart . . . ," "I Left My Heart in San Francisco . . . ,"
"Heart and Soul."

More than a hundred transplant recipients showed
up. They were easily identified by the spangled red hearts
pinned to their dress straps and suit lapels. Columbia-
Presbyterian has done so many transplants by now that
the party organizers who were handing out the spangled
hearts at the door actually had to ask people as they came
in whether or not they'd had a transplant. Upon hearing
the question, each "have not," almost to a person, clutched
a hand over his or her natural heart as though to make
absolutely sure no one had taken it.

Seated at one table was a sixty-five-year-old man named

Seamus Healy, a runner-up for the Irish Olympic Bicy-cling Team back in 1955. "My old heart was mush," he shouted at me above the band. "I got a young teenage girl's heart now. I pray for her every morning. She's my angel. How can I forget her?"

All the recipients I met that night spoke in the same rev-erent tones about the angel in their chests, about this gift, this responsibility they now bore, and the little prayers they said to this other person inside them. They talked of whole new sensory responses, cravings, and habits. It was as though I were meeting the members of some strange new cult, the tribe of the transplanted.

Some researchers have explained the phenomenon in terms of the "surprised heart theory": the donors' deaths being so sudden that their hearts continue to act as if they were still in their original owner's body. Others ascribe it either to the heart's neurochemicals, sometimes called the "little brain in the heart" theory, or to neuropeptides, strings of amino acids, some of them found to reside in the heart. They pass along information by locking on to matching cellular receptors and then stimulating an elec-trical charge in the neurons.

"All I know is I had a talk with her the night after my surgery," a fifty-three-year-old woman named Betty Dio-taiuti said of her new heart and its donor. "I told her, 'I hope you're not an insomniac.'"

Near the end of the evening, all the spangled-heart wearers were brought together at the front of the room for a group photo. All eyes turned their way. I noticed here and there above the tables the telemetry poles of the Status 1 patients who'd been allowed off their floor for the occasion, all of them staring up toward the front of the room as well. The wife of the patient just beside me was crying, rubbing her husband's arm, telling him not to worry, that he'd be getting his new heart soon.

Is it even possible, I wondered as our Organ Unit sedan began to slow along the turnpike, to anthropomorphize a heart? What, after all, is more human? Ours, the one we had in the trunk, was merely about to be introduced into a new bloodstream and brain, into which it would, in turn, invariably spill the residual whispers of its first life's conversation.

The sirens went suddenly silent. I looked up, thinking we must be at the hospital, but saw instead a row of turnpike toll booths, our driver John Bezares pulling to a stop at one of the "cash only" lanes. It seemed to me a bit odd, given our cargo.

"Are you kiddin'?" Bezares shouted back as we sped off again. "Five-hundred-dollar fine!"

Chapter 19

MY NIGHT WAS SUPPOSED TO END WITH THE heart's delivery. My "official clearance" extended no further. We rolled the cooler up to the doors of Operating Room No. 22. Dr. Slater pushed them open. Dr. Rosen entered with the cooler. I remained for a moment in the doorway, trying to get a quick look at the proceedings inside. It was a scene not unlike the one I'd just left, the same light-blanched, blue-and-white tableau of a patient lying on the table, mostly covered, surgeons, and various OR attendants positioned all around. I watched all their heads lift and turn with the cooler's entrance.

"It was a beautiful arrest," Dr. Rosen announced, rolling the cooler off to the side of the operating table.

The lead surgeon's headlamp was glaring in my eyes. I raised a hand to block it, hearing all the while the nervous whispering of some of the OR attendants: "He's not supposed to be here. He's not supposed . . ."

"Charles," a familiar voice announced, "nice to have you along. How about scrubbing for surgery."

I stepped inside, saw first a pair of neon-green clogs and then the unmistakable outline of Dr. Michler. He'd already set back to work upon the patient. An attendant approached and escorted me across the room to the prep area behind another set of doors on the far side of the OR.

"You're quite a lucky guy," Dr. Rosen said to me in the prep room, he and Dr. Slater washing their hands at adjacent sinks before heading toward the door leading into the surgeon's locker room, where they would change back into their civilian clothes. "You get three or four more hours of surgery."

Once scrubbed, I reentered the OR, hands aloft. Following Dr. Michler's signals, I found myself a spot midway along the table, directly opposite him. I had to wedge myself in beside two other medical personnel, who barely moved. They didn't seem pleased about my being there.

"You can drop your hands, Charles," Dr. Michler said in that same airy voice. "At your sides, just off the table is fine."

The patient's head was, as with the donor's, covered. I knew only what Dr. Rosen had told me on the way to the harvest: that he was a man in his mid-fifties, a Status 2 patient, dying of heart failure. His chest was the only un-covered part of him. His breastbone had been sawed open and pried apart with heavy metal clamps, a badly distended heart occupying nearly all of the chest cavity. It seemed to balloon more than beat.

I looked back toward the covered head, and then, furtively, around at the surrounding tables of instruments, trying to see if I could spot a patient's biographical chart like the one I'd seen for the donor back in the harvest room. Somehow his anonymity seemed less essential to me than hers had. It was, in fact, for her sake now and that of her still-packaged, icebound heart that I most wanted to know his identity, to know who it was through whom she was about to continue. But there was no chart in sight, and I soon found myself—as I have in all of our subsequent visits at the dream-fountain in the 4:00 A.M. hour—assigning him my father's visage.

I watched Dr. Michler insert a cannula now—essentially a small, tubelike valve—near the base of the patient's vena cava. To the top of the cannula he then attached the lines from the heart-lung bypass machine and flipped a switch. We all watched as the patient's entire bloodstream drained down into the machine, its circular console windows awash now in red.

Dr. Michler next placed a clamp on the aorta, then cut it and the remaining vessels to which the heart was moored. He then reached in with both hands and, just as Dr. Rosen had with the donor's healthy heart, he lifted out the dying one, a flaccid, nearly shapeless pink-and-white bag.

"Do you see," Dr. Michler said, "how sick it is?"

He brought the heart over to a side table and placed it in a metal bowl. I watched as the attendant beside him went to the picnic cooler and undid the bags containing our package. He brought the bags over to Michler, who reached in and took out the heart. He washed it in a bowl of ice-cold saline, then walked it back to the operating table. He was about to set the heart into the patient's chest when he paused, turned, and went back to the side table, placing the heart back in the bowl of saline alongside the one he'd just extracted.

"Get the camera," he told an attendant. "This will make a great picture for my lectures."

I held my place at the table, staring down at a hollowed-out chest, an abandoned nest, this man's bloodstream

entirely bypassing its only previously known banks, his existence's abeyance now the complete obverse of the donor's: a
fully functioning brain above an empty parlor, his mind's
lifelong conversant gone, its only respondent now a thrumming machine.

I stood there remembering how badly, in those waning
days of my father's life, I'd begun to want a fully mechanical
heart for him, hoping he'd be able to hang on long enough
for one of the models that were then just in the early devel-
opmental stages. The first experimental implantations
didn't occur until four years after he'd died, and still I found
myself following the plight of each of the recipients as
though they were part of a great pioneer mission to Mars.

Especially the story of William Schroeder, the second
man in history to receive a totally artificial heart, and the
one who traveled the farthest with it. The first recipient,
Dr. Barney Clark, a dentist, lived 112 days; Murray Haydon,
488; Jack Burcham, 10; and Leif Steinberg of Sweden, 210.
Schroeder would live for 620, close to two years, the nearest we've ever come to the dream of a life without our
native hearts, with a perfectly timed, infinitely replicable,
and reparable power source.

It is all a nearly forgotten foray now, but one that was
imbued at the outset with such hope and expectation, and
not merely for the ailing recipients. This was to be the true
apotheosis of the pump, the Jarvik-7, that would one day
obviate our natural, mortal hearts, "sick with desire, and
tethered to a dying animal." They might be cut out of the
conversation for good now, in order to make way for a

heart and a life ongoing: that climb we have long dreamed of making up into "the artifice of eternity."

If only my father had held on a few more years, I kept thinking as Schroeder's journey continued, he'd be getting one of his own Jarvik-7's, and, some years later, an even newer model with all the kinks worked out. The very one that I'd be getting when my natural heart gave out, the two of us, a pioneer father-son duo of men after their own hearts, with all the time in the world now for that never-to-be-had rapprochement, sitting together in some suburban solarium, chatting on and on above the whooshes and clicks of our mechanical pumps, about the wonders of modern technology and all of the anonymous little electronic parts that went into it.

I don't remember when exactly it dawned on me that things were going irreparably wrong for William Schroeder. I do recall one photograph of him, however, taken some five months after his surgery. He's outside, in a wheelchair, going on a fishing trip with his family. One son is pushing him up a hill, while a daughter walks alongside with the bookbag-sized portable heart-drive slung over her shoulder and, in an upheld hand, the drive lines through which flows the air that powers her father's Jarvik-7 heart.

Everyone is smiling eagerly for the camera, while Schroeder, in the foreground, sits slumped forward and tilted to one side, his right hand tugging at his oversize T-shirt near where the drive lines enter through his stomach, his eyes fixed in a downward gaze. It's an unsettled posture, poised somewhere between protest and powerlessness, and his expression, I thought, the longer I looked at it, could have no "like," no simile: It's the look of a man who has lost his heart.

None of the Jarvik-7 recipients took notes or kept journals, too sick with strokes, infections, and high fevers that left them for most of the time in a twilight state. The only record of what it was like to live without a human heart is contained in the reams of technical data doctors collected, one medical film of Schroeder's operation and recovery, and the countless newspaper and magazine stories that, even at the time, began to read to me less like the stuff of bold breakthrough than like a chapter out of cheap sci-fi.

Dr. Robert Jarvik, president of Symbion, the company that manufactured the plastic-and-titanium Jarvik-7, was more an inventor than a medical man, having gone directly from medical school into the development of artificial organs. Schroeder was the first human patient he'd treated. Medstar was the company filming Schroeder's experience, Black Star the photography agency. In one issue of the *Journal of the American Medical Association* that I found, Dr. William De Vries, the surgeon who implanted Schroeder's Jarvik-7, lists the chronological order of events leading up to the surgery in the manner of a space-launch countdown—minus twelve days, then minus eight, right up to the moment of severance and implant.

The Jarvik-7 did prove to be the ultimate pump, never missing a beat in any of the recipients, the only mishap being a broken valve in Barney Clark's heart that was promptly repaired. And yet, even on the purely physical level, it failed to hold up its end of that conversation our hearts have with the rest of our organs. All recipients, for example, suffered acute kidney failure, even though with the turn of a switch on the Utahdrive—the 230-pound, shopping-cart-size machine that powered the Jarvik-7—a perfect blood pressure and heart rate was maintained. It's as though our organs ask for more than an efficient, robotic output, ask for that subtlety and variety of pulse exhibited by a natural heart. Trying to live with a Jarvik-7 was, in essence, akin to an orchestra trying to perform a fugue under the leadership of a tone-deaf, monorhythmic maestro.

The doctors tending to the artificial-heart recipients had little time to consider its psychological and emotional effects—constantly scrambling as they were to deal with all the riotous physical responses to it. Newspaper and magazine articles I read would mention in passing the deep depression patients suffered, but this was routinely ascribed to the many physical setbacks they kept experiencing. Mostly the talk was of progress.

There was, as with any pioneering mission, the series of notable firsts: Schroeder's first beer with his new heart; his first birthday (I remember this because the only living man on the planet without a natural heart at that time happened to be born on St. Valentine's Day); and, four days after that, his first trip outdoors, hooked up to Heimes, the portable heart-driver. On the way back from that particular outing, Schroeder was wheeled up to the door of the room where Murray Haydon, who'd just received his Jarvik-7, was lying in bed. It was the first meeting ever of two men with artificial hearts.

There is a picture of the event. I have it still, taken by Black Star photographer William Strode. It is shot from behind Haydon's left shoulder, so that only the side of his face is visible. Schroeder, as in the the first fishing-trip photo, is seen slumped sideways in his wheelchair, and there's the same marked difference between his expression—a distorted, nearly centrifugal ache that seems to show up more and more in later photos—and the eager smiles of the nurses, doctors, and family members around him. His right hand is held up in a limp wave, almost a plea, to Haydon, who's holding up his left hand in precisely the same manner so that it appears each man is staring, bewildered, at his own reflection.

In the one psychiatric study I found on an artificial-heart recipient, Barney Clark, the doctors expressed concern about Clark's mental condition, the fact that he "experienced periods of despondency and asked to die or be killed." Clark's wife described him as having suffered a significant loss of personality, of being like "a wall." Margaret Schroeder made similar remarks to the effect that her husband didn't seem himself and was often barely willing to speak, was weepy and depressed.

What it felt like to live with a Jarvik-7 is a secret the recipients have long since taken with them. But I've always wondered if the depression and the loss of aspect in their personalities had also to do with a proportionate loss of aspect in their new hearts' responses to the varying emotional stimuli around them. Could their depression have stemmed not only from the physical suffering, but also from the now thorough severance they were experiencing of that conversation between the heart and brain? A disparity the recipients recognized, say, between the way their natural heart once reacted when they saw their wives enter a room and the way their new one couldn't in that same instance.

Schroeder went on a fishing trip. He also, despite his fatigue and feverish delirium, attended a minor-league baseball game. If, I asked myself, his body had completely accepted the Jarvik-7, and no clots had formed to cause strokes, and no infections to bring on high fevers; if there had been no physical side effects mitigating perception and he had felt perfectly fine and lucid, what would his perceptions of and responses to these outings have been like? If the team he roots for mounts a late-inning rally and his heart doesn't rally with them, wouldn't that feel different? Or if, at the moment a fish is struck and brought to the surface, his new heart does not race and rise with it, is he aware of that?

The Jarvik-7's recipients did, of course, feel, and deeply. Schroeder cried at his son's wedding, held in a hospital chapel. He cried continually over his predicament. But when someone's heart is no longer working in concert with those feelings, does he feel that and cry more? Having had his heart cut out of that lifelong internal to-and-fro, does he begin to dwell more exclusively in the brain; to merely draw, in computerlike fashion, on the memory of his old heart's role in the emotional equation, and thus proceed to a place for which there is no human precedent: a person whose heart is all constancy, all riverbank and no rainstorm.

Schroeder had been gasping for every breath the day in November of 1984 when his spent heart was replaced. For the first eighteen days, the Jarvik-7 pumping away with vigor and efficiency, Schroeder was breathing freely again, feeling restored. On the nineteenth day he suffered a stroke that might have killed a man with a normal heart. But the Jarvik-7, oblivious of the surrounding turmoil, continued vigorously pumping blood to the damaged area of the brain, and thus raised a philosophical question its creators hadn't foreseen: Considering the suffering Schroeder was to endure, did the Jarvik-7 enable him a supernatural recovery or deprive him of a natural death?

I remember the dizzying promise of those first two and a half weeks after the implant, Schroeder wheeling around the hospital halls, making people put their hands to his chest to feel what he kept saying felt like "an old-time threshin' machine." He spoke on the phone with President Ronald Reagan, asking him to locate a missing social security check; wrote a letter wishing "best of luck" to a man who'd just received an artificial ear (also made by Symbion); and conducted a number of interviews with the press. One interviewer asked Schroeder if he felt a piece of him had died when he lost his own heart.

"No," Schroeder answered. "It's of no value to me. I've

got a new one. It just thumps away in there. And I can feel
it. It doesn't bother me at all."

But months later and despite his sometimes clouded
mental state, he was heard to express very definitive anger
and frustration over his new heart, and for the most basic
reason: the noise and the looming presence of his heart's
drive machine, an intrusion that would soon become all
too apparent when he and his wife moved into the quiet,
specially equipped apartment across the street from the
hospital, lording over their only private time together.

Margaret Schroeder would say afterwards that throughout her husband's ordeal, "she felt she was a prisoner of the
artificial heart." In the basement of that special apartment
were the generators and backup generators for the heart's
drive machine, all of which, had the Schroeders been able to
return to their normal life, would have filled the basement
of their home in rural Indiana: an entire house thrumming
against the cold of a Midwestern winter night to power the
beating of one man's heart.

A professor of health law and an authority on patient's
rights visited Schroeder one day in his private apartment.
When he was asked how he liked his "machine," Schroeder
said over and over that he hated it. He could find, in the end,
no way of situating himself around a heart half outside of
him, was unable to get accustomed to a shape and a noise
that he could neither contain nor forget, thus depriving him
of perhaps the most simple, individuating moment we have
in our lives, when, at rest, we curl up, involute, wombward,

around our own body's original inner thrum; curl up the way I remember my mother describing my father doing the very last night of his life, around both her and himself, as though afraid to let go.

It was at the very end of Schroeder's life, barely able to speak because of another stroke, that he tried to signal back to his wife from wherever it was the Jarvik-7 had taken him. He'd point to himself and then to her. Point to the heart-drive machine alongside him, and then to the empty space beside her. He kept repeating the signals and asking, "Why?" Confused at first, his wife finally asked him if he meant he wanted her to have a heart machine, too. Schroeder, nodding vigorously, whispered, "Yes!"

A camera flashed. I looked up to see Dr. Michler waving me over. I went over and stood looking down at the table before him, focusing first upon the metal bowl with the healthy heart, a bright, multicolored reticulum of veins and arteries clinging to its deep magenta core; and then on the bowl with the sick one, which—I had to look again to be sure— was still moving in its shallow pool of blood: slight, swelling heaves, like the slow-gaping maw of a dying beached fish.

"You see that?" Michler said, eyes wide. "It's still struggling for life."

My father was asleep that last morning, near dawn, February 1, 1980. On the night table beside the bed was his Lawrence Sanders novel, The Sixth Commandment, the bookmark placed at page 313. I've scoured it many times

since for retrievable portent. Never found any. These, anyway, are the last words that might possibly have passed that
night before his closing eyes:

> We finished our coffee. I thanked Sam Livingston
> and left. He didn't rise to see me out. Just waved a
> hand slowly. When I closed the door, he was still
> seated at the table with empty cups and a full ash-
> tray. He was an old, old man trying vainly to recall
> a dim time of passion and resolve.
>
> When I got up to the lobby, the desk clerk
> motioned . . .

On the newly upholstered settee at the foot of my parents' bed was an open suitcase packed for the train trip
they were to take together from Chicago to St. Louis to
visit with the nearest my father had to living relatives. He'd
been on a number of similar missions in the previous
months, uncharacteristic meanderings from his usually
stringent sales-trip rounds, the man committed to making
good time now making a point of flouting it, wanting only
to visit with old friends.

Definite portent, that and his exceptional clinginess all
through that last night, never wanting to let go, "as though
we were kids again," my mother said, and I've often
thought since of the letters that I found in that box as well
on my train trip to Washington, D.C., letters he'd written
her during their courtship in the winter and spring of
1947, love-mad, over-the-top missives, the script marked by

the same bold stanchions and elegant connective spans, a number of the letters delivered by him personally, having been composed just a half hour before he was to meet my mother in Manhattan or downtown Brooklyn for lunch.

He must have been holding on that last night, I've often thought, in such a way as to make both her and his blood's circuit that much shorter around his own heart's inward failing. Later that morning, in a last bit of predawn darkness, he would sit up straight in bed, and say only the words "Oh no," because he already knew he'd been abandoned. That he was too far out and his heart no longer there along with him, and the water rising, with nowhere else to go.

You go blindly, furiously, toward any suggestion of an opening. Room after narrowing cave room, closing in, and the voices going silent, until the walls eventually meld up ahead and drop beneath the now chest-high waterline, with only the faintest, wobbling tendrils of submerged sunlight signaling a way out.

I wonder now if he saw it all even then, that kid who began screaming behind me, the most foulmouthed and feared in camp, the one who had to be shepherded to the back of the line and then later pulled out by a rope. If, perhaps, the whole picture didn't briefly coalesce before him right there in the cave that morning, the scenario in which they, the adults, don't, in fact, have a way out; and the carnival balloon man does finally drift off and his hand come lose, and all of his carefully tied-off wares float free.

I remember trying to take one of those deep yet nerve-begrudged inhales, then ducking under for the frenzied, wall-grabbing swim, that faint call to stay playing along my ear's inner gill, and then the waiting hands pulling me up into the otherness of air and the news of him: a feeling of both deliverance and displacement, that unshakable suspicion, yet again, that I am not supposed to be here.

"Hold this!"

I stood now looking at a pair of surgical scissors, Dr. Michler handing them to me, keeping me in the game.

"Scissors?" he ordered.

I handed them back over, watched as he trimmed the harvested heart along the top so that its parts would meld better with their corresponding connecting points in the recipient: the cuffs of the old heart's upper chambers, the apertures of the aorta, and of the pulmonary artery. Then he began madly sewing together all the vital connections: the aorta, the vena cava, the pulmonary artery and veins.

At 8:35 A.M., there was a marked hush in OR No. 22. "This," Michler said, "is the moment of truth."

He reached down, closed off the cannula, and removed the aortal clamp, the gesture by which the bypassed blood would begin returning to its new and yet primordially familiar home.

We all stood in silence. For minutes, nothing happened. I nervously recalled what Dr. Slater had told me in the course of the drive back from Newark. "It's no small thing we do to these hearts," he said. "We stop them, hold them like a baseball, freeze and pack them. Sometimes they just don't start up again."

I kept thinking of the warm blood now, sloshing up

against the iced muscle walls. Kept staring at the heart and, ridiculously, at its new owner's covered face, as though something of the turmoil that was now going on within his body might first register there. As though with the sudden onslaught of healthy heartbeats would also come instant animation: this man sitting up there before me just as my father had in bed that morning—only now in the direction of life—the blue towel falling away, his face, finally, revealed.

Dr. Michler reached in and gave the heart some gentle, coaxing squeezes. I didn't think such a thing was even allowed at this point. For all I'd been through that night with this heart—the arrest, the excision, and the packing in ice; the delivery and unpacking, the last-minute trimming, and the implantation—now that it was back in someone's chest, it seemed to me once again something off limits, precious, other.

Dr. Michler let go. Reached in again. Let go. Still nothing. How deep, I began to wonder, is a cellular memory? How much time out of its element is too long for even a heart to remember what to do?

And then, with the same slow lurchings of that sick, excised heart in the metal bowl on the side table—only now in the direction of life—it did. Came alive in fits and starts, beat erratically for a time, then took off, pumping at 90, 100, as high as 145 beats a minute before coming back down into the high nineties, the man's blood pressure all the while remaining disconcertingly low.

"He's so dilated," Michler kept saying, the recipient's veins having been opened so wide for so long in want of

269

blood from his sick heart that they were now not giving back readily enough the blood the new heart was giving, not maintaining that proper inner pressure gradient at which our lives are ever poised.

The heart was fiery red by now, "innate heat," a primordial star throbbing away in its own universe of being. In each heart, yours, mine, is the story of the very first heart, a story we are tracing now all the way back to its very first beats, the new explorers and mapmakers completing in reverse that wall of fetuses that I stood before so many years ago, the invisible panes, both of our own individual beginning and, within the airy frames of that backwardly disappearing diptych, of all life's beginning.

Day Fifteen: Our heart's first, tentative, tubular beats. The commencement of the "vitelline circulation." The day that our troubles truly begin to begin.

Day Twelve: No discernible shapes at all, no apparent articulation of the embryo or its "inspired" parts, the ovum's replicating cell layers just beginning their endless inward looping and pinching off to form the embryonic axis, that sinus or inner groove known as the "primitive streak," about which our lives begin to take shape.

Day Six: The heart, the rudimentary heart, has yet—like the proto-brain that it starts out nestled beside—to even cohere, its two formative halves only beginning their journey from opposite sides of the body toward the median line where they will coalesce, fuse.

Our hearts, it will surprise no one, are born broken. They are not, as was long ago and logically assumed, the first of us to form, "the first living and the last dying" as the French battlefield surgeon Ambroise Paré wrote as recently as the sixteenth century.

We dwell, in fact, for the first two weeks of our existence in a kind of heartless abeyance, a brief and faintly distant echo of the vast expanse of time evolution took to even invent the heart; the overwhelming majority of the Earth's history, a period during which the planet teemed with life and yet was utterly heartless. A one-note planet: vast mats of the same fully independent, self-replicating, single-cell bacteria, forged in the fires of the Earth's early formation, living on carbon monoxide and other poisonous gases, expelling as waste the tiny puffs of oxygen that over billions of years would eventually build the atmosphere that allowed the formation of those very first oxygenated cells that would one day rove over into two, and from there into infinite variations of complexity.

We contain, it turns out, that moment still within each one of our cells. Advanced fossil studies and DNA analysis of the single-cell anaerobic bacteria that live today in boiling hot, deep ocean vents known as fumaroles, show that the different working components of the cells in our bodies, known as organelles—the nucleus, the mitochondria, and so on—are themselves the distant descendants of those early single-cell bacteria from the days of the Earth's fiery formation and eventual cooling. Hell's fires, heaven's airy vapors. These, we now know, aren't imaginative conceits. They're deep cellular memories.

In each heart's beginning, yours, mine, is the beginning of that very first heart, an event that, "in biological time," occurred only yesterday, a moment we have no need to recall, because we are its consequence. Our hearts begin the way life itself first did: as an inkling, no more, a noiseless cellular nudge, like the first silent stroke of the maestro's wand, that distinct nether beat before the initial note of what soon builds into a grand symphony, a symphony to be set, as it is with all of life's complex arrangements, in the key of the heart.

One cell, one note, nudges into another, and the next and then a line of notes, which, in turn, roves over into another line, and the next, and now these three begin to turn back inward on themselves, connecting up, end to end, pinching off, outlining the myriad motifs that will be articulated later: the brain, the nerves, the veins, and the limbs; the liver and the spleen, and that most supple, overarching theme of the skin.

But we are only notes at first, you and me, swirls of self-driven, cellular notes, drawn from billions of years of accumulated biological scores, the various themes of which continue to build within us for hours, days, weeks, without as yet one discernible beat behind them, the heart, like a late-arriving maestro, hurrying to compose itself in order to conduct the remainder of the very composition of which it is the formative part.

Day One: The fertilized ovum. All potential. Elliptical urge. One cell containing the whole of biological time. Within each conception—yours, mine—is the echo of all life's inception, an event so distant, so out of time, it can only be recalled now by heart: one irreversible moment in the wake of our universe's formation when, light having at last decoupled from matter, the fiery-tailed seed of a carbonaceous comet penetrated the gaseous ovum of a newly formed Earth (the only viable egg in our solar system's sac), somehow triggering life's switch.

"Now that's a bit more like it."

Dr. Michler stood there, looking up at the heart monitor, the medications he'd ordered beginning to right the blood pressure now. He then instructed his assistants to attach a tiny wire electrode to the heart muscle and set the heart rate at ninety-nine beats a minute. It would be kept there for the coming days to help train the heart, suggest to it the rhythm that it would have to eventually assume on its own.

I looked over at the clock. It was nearly eleven-thirty in the morning. Dr. Michler's headlamp was off. I could feel myself just then beginning to come down off that crest of fast, shallow breathing that I'd been riding for so many hours. Our donor's heart had resumed its life now within another person, someone who, despite having never seen his face, I felt I'd already come to know as intimately as one can another human being.

And then—for reasons I'll never understand and was later asked by Dr. Michler in the surgeon's locker room not, for legal reasons, to explore further—one of the assisting surgeons who'd been standing beside me throughout the procedure and seemed the most unsettled by my presence there that morning, took hold of my right arm just above the wrist and placed my hand directly upon the beating heart.

I expected something softer, a kind of gelatinous warble. But there is nothing frail or hesitant about the separate life that lives within you. I'd fully met it now, and so my own, and my father's, and all hearts. It's a force above which thoughts and words disperse and drift like the vapor above a sea-swallowed lava flow.

I can feel it still: the hot, swelling ease of diastole, followed hard by systole's stiff, pounding blow, like the rigorous, full-length press of a just-landed fish, all of its cells rebelling in unison against so coarse and empty a medium as air.

About the Author

CHARLES SIEBERT is the author of *Wickerby: An Urban Pastoral* and the novel *Angus*. His essays, articles, and poems have appeared in numerous publications including *The New York Times Magazine*, *The New Yorker*, *Harper's*, *Esquire*, *Vanity Fair*, *Outside*, and *Men's Journal*. He lives with his wife in Brooklyn, New York.